The Decomposition of Figures into Smaller Parts

Popular Lectures in Mathematics

Survey of Recent East European Mathematical Literature

A project conducted by
Izaak Wirszup,
Department of Mathematics,
the University of Chicago,
under a grant from the
National Science Foundation

Vladimir Grigor'evich
Boltyanskii
and
Izrail' Tsudikovich
Gohberg

The Decomposition of Figures into Smaller Parts

Translated and
adapted from the
Russian edition by
Henry Christoffers and
Thomas P. Branson

The
University of Chicago
Press
Chicago and
London

The University of Chicago Press, Chicago 60637
The University of Chicago Press, Ltd., London

Printed in the United States of America
84 83 82 81 80 5 4 3 2 1

Library of Congress Cataloging in Publication Data

Boltianskiĭ, Vladimir Grigor'evich.
The decomposition of figures into smaller parts.

(Popular lectures in mathematics) (Survey of recent East European mathematical literature)
Translation of Razbienie figur na men'shie chasti.
1. Combinatorial geometry. I. Gokhberg, Izrail' TSudikovich, joint author. II. Title.
QA167.B6513 516'.13 79-10382
ISBN 0-226-06357-7

Contents

Preface

This book is devoted to some interrelated problems of a new, rapidly developing branch of mathematics called *combinatorial geometry*. Common to all the problems examined here is the notion of "cutting" a geometric figure into several "smaller pieces." There are several different criteria for what constitutes a "smaller piece"; hence this book necessarily treats several different problems. All the theorems proved here are very recent; the oldest of them was proved by the Polish mathematician Karol Borsuk about forty years ago. This theorem of Borsuk is the core around which all of the subsequent exposition unfolds. The most recent theorem is barely a year old.

The topics treated in this book are well within the grasp of bright and interested high school students. At the same time, the book introduces the reader to a number of the unsolved problems of geometry.

This family of problems is the subject of another book by the same authors, *Theorems and Problems in Combinatorial Geometry* (Nauka, 1965). That book, however, deals chiefly with problems of three-dimensional and higher-dimensional spaces. The present book concerns itself only with problems of plane geometry, and can thus be used by high school mathematics clubs. *Theorems and Problems in Combinatorial Geometry* will be useful, however, to readers interested in continuing further.[1]

The remarks at the end of the book are intended for the more advanced reader.

1. Other references on this subject are Hugo Hadwiger, Hans Debrunner, and Victor Klee, *Combinatorial Geometry in the Plane* (New York: Holt, Rinehart, and Winston, 1964); Ludwig Danzer, Branko Grunbaum, and Victor Klee, "Helly's Theorem and Its Relatives," in *Convexity*, Proceedings of Symposia in Pure Mathematics, vol. 7 (Providence: American Mathematical Society, 1963), pp. 101–80.

1 Division of Figures into Pieces of Smaller Diameter

1.1. The Diameter of a Figure

Consider a circular disk of diameter d in the plane. The distance between any two points M and N of this disk (fig. 1.1) cannot exceed d. At the same time, it is possible to find points A and B in the disk which are separated by a distance of exactly d.

The question naturally arises: Can we assign "diameters" to plane figures other than circular disks? The above comments suggest the possibility of defining the diameter of a plane figure as the greatest of the distances between pairs of its points, if such a greatest distance is attained (fig. 1.2). For each of the figures considered in this book, such a maximal distance will exist, and thus each figure will have a well-defined diameter (see remark 1).

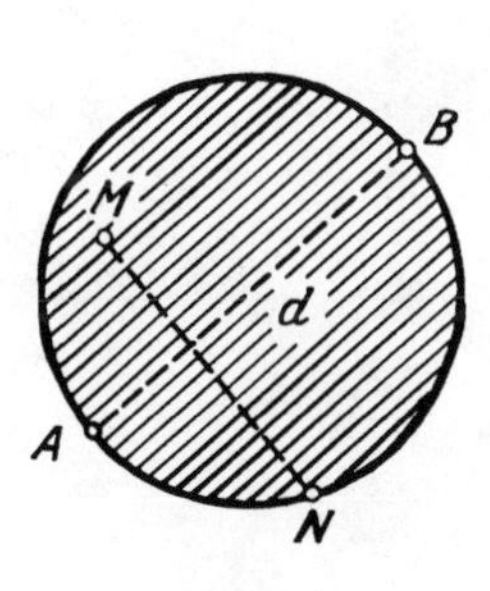

Fig. 1.1

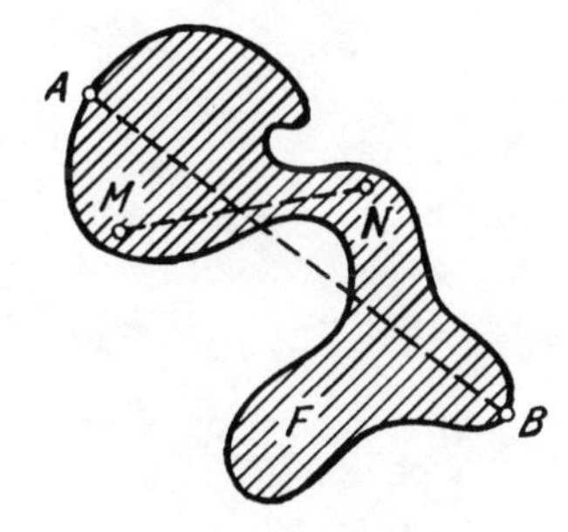

Fig. 1.2

Suppose, for example, that the figure F is a semicircular region (fig. 1.3). Let A and B denote the endpoints of the semicircle that forms an edge of the figure. Then it is clear that the diameter of the figure F is the

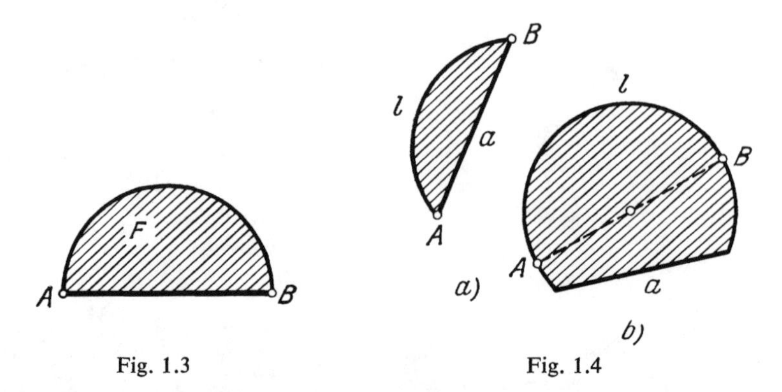

Fig. 1.3

Fig. 1.4

length of the segment AB. More generally, let F be a segment of a circular disk bounded by an arc l and a chord AB of length a. Then if the angle measure of the arc l is not greater than the semicircle's 180° (fig. 1.4a), the diameter of F is a; if the measure of arc l is greater than 180° (fig. 1.4b), the diameter of F coincides with the diameter of the whole disk. It is easy to see that if F is a polygon (fig. 1.5), then the diameter of F is the largest of the distances between its vertices (see remark 2). In particular, the diameter of any triangle (fig. 1.6) is the length of its longest side.

We note that a figure of diameter d may contain more than one pair of points separated by the distance d. For example, a noncircular ellipse (fig. 1.7) has only one such pair of points; a square (fig. 1.8)

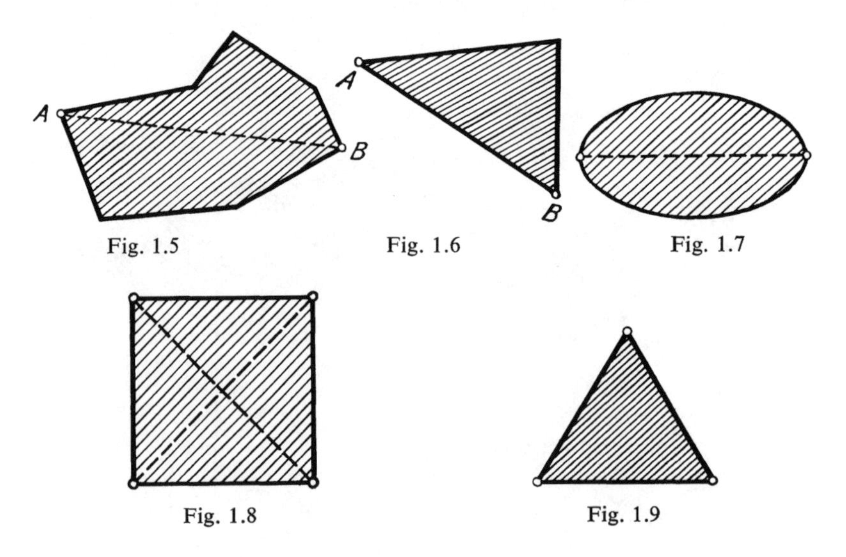

Fig. 1.5

Fig. 1.6

Fig. 1.7

Fig. 1.8

Fig. 1.9

has two; an equilateral triangle (fig. 1.9) has three; and, finally, a circular disk contains infinitely many such pairs.

1.2. Formulation of the Problem

It is not hard to see that if a circular disk of diameter d is cut into two parts by some path MN (where we consider the path to lie in both of the new sets), at least one of these parts will have the same diameter d. Indeed, if M' is the point diametrically opposite M, then M' must belong to one of the pieces, and this piece (containing the points M and M') will have diameter d (figs. 1.10*a* and 1.10*b* (see remark 3)). Nevertheless, a circular disk clearly can be cut into three pieces, each of which has diameter less than d (figs. 1.11*a* and 1.11*b*).

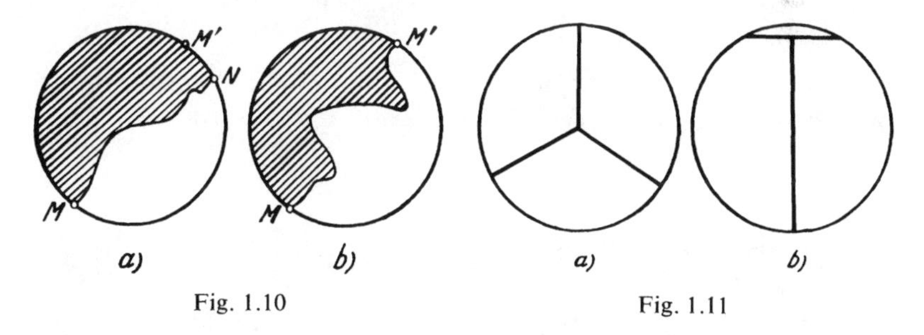

Fig. 1.10 Fig. 1.11

An equilateral triangle with side d possesses the same property (if it is divided into two pieces, one of the pieces must contain two vertices of the triangle, and the diameter of this piece will be d). There are, however, figures that can be divided into two pieces of smaller diameter (figs. 1.12*a* and 1.12*b*).

We shall consider the general problem of dividing a given bounded figure F into pieces of smaller diameter (see remark 4). The smallest number of pieces necessary will be denoted by $a(F)$. Thus, if F is a circular disk or an equilateral triangle, $a(F) = 3$; if F is a noncircular ellipse or a parallelogram, $a(F) = 2$.

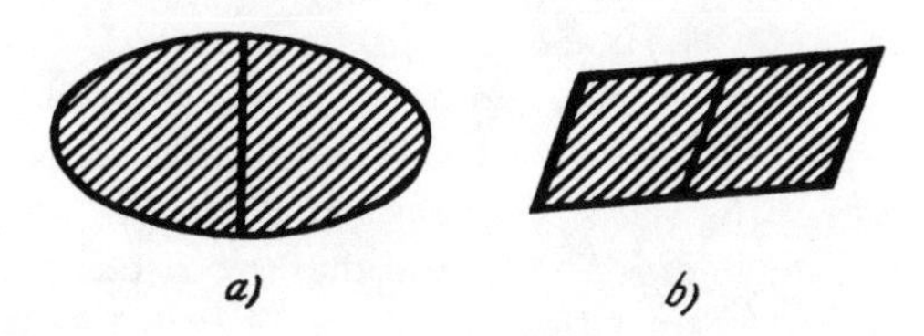

Fig. 1.12

The question of what values $a(F)$ can assume was posed and solved in 1933 by the noted Polish mathematician Karol Borsuk (see remark 5).

1.3. Borsuk's Theorem

We have seen examples of plane figures for which $a(F)$ takes on the values 2 and 3. There naturally arises the question of whether one can find a plane figure F for which $a(F) > 3$—that is, a figure for which we cannot manage with three pieces, but need four or more (to divide it into pieces of lesser diameter). It turns out, in fact, that three pieces are always enough; this is the assertion of the following theorem, established by Borsuk:

THEOREM 1.1 (Borsuk's theorem). *Any plane figure of diameter d can be divided into three pieces of diameter less than d; that is,* $a(F) \leq 3$.

The crux of the proof lies in the following lemma, which was obtained in 1920 by the Hungarian mathematician Pál:

LEMMA 1.1. *Any plane figure of diameter d can be enclosed in a regular hexagon whose parallel sides are separated by a distance of d* (fig. 1.13).

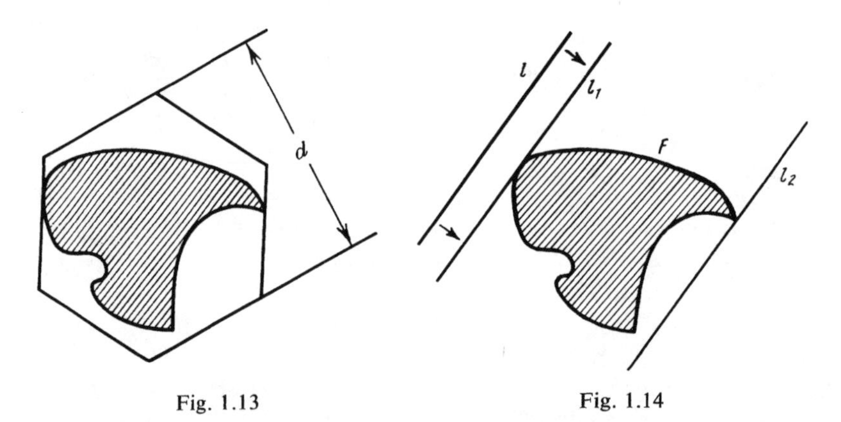

Fig. 1.13 Fig. 1.14

Proof. Take a line l not intersecting the figure F, and move this line closer to F, keeping it parallel to its original position, until it touches the figure F (fig. 1.14). The line l_1 thus obtained has at least one point in common with F, and all of F lies on one side of l_1. Such a line is called a *supporting line* of (or to) the figure F. Now draw a second supporting line l_2, parallel to l_1 (fig. 1.14). It is evident that all of F will lie within the strip between the lines l_1 and l_2, and that the distance between these lines will not exceed d (since the diameter of F is d) (see remark 6).

Now draw two parallel supporting lines to F, m_1 and m_2, at a 60° angle to l_1 (fig. 1.15). The lines l_1, l_2, m_1, and m_2 form a parallelogram $ABCD$ with an interior angle of 60° and altitudes no greater than d which completely encloses the figure F (fig. 1.15).

Now draw two parallel supporting lines p_1 and p_2 to F, forming an angle of 120° with l_1, and denote by M and N, respectively, the bases of the perpendiculars to these lines from A to C, the endpoints of the long diagonal of the parallelogram (fig. 1.15). We shall show that the direction of l_1 can be chosen in such a way that $AM = CN$. Indeed, suppose that $AM \neq CN$; without loss of generality, say $AM < CN$. Then the quantity $y = AM - CN$ is negative. Now we continuously change the direction of the line l_1 until it is turned around 180° (leaving the figure F fixed). The other lines, l_2, m_1, m_2, p_1, and p_2, will also change their positions, which are completely determined by the choice of l_1. Therefore, as the line l_1 is moved around, the points A, C, M, and N will also be shifted continuously (see remark 7), and so the quantity $y = AM - CN$ will change continuously. But when the line l_1 is turned 180° while remaining a supporting line, it must occupy the position previously occupied by l_2. Therefore we obtain *the same* parallelogram as in figure 1.15, but the points A and C, and likewise M and N, will exchange roles. Consequently, in this position y will be *positive*. If we graph the continuous change in y as a function of the angle through which the line l_1 is turned (fig. 1.16), we see that there must be a position of l_1 for which *y vanishes*; that is, for which $AM = CN$ (since y, while changing continuously from a negative value to a positive value, must become zero somewhere). We examine the positions of all our lines at the orientation of l for which y vanishes

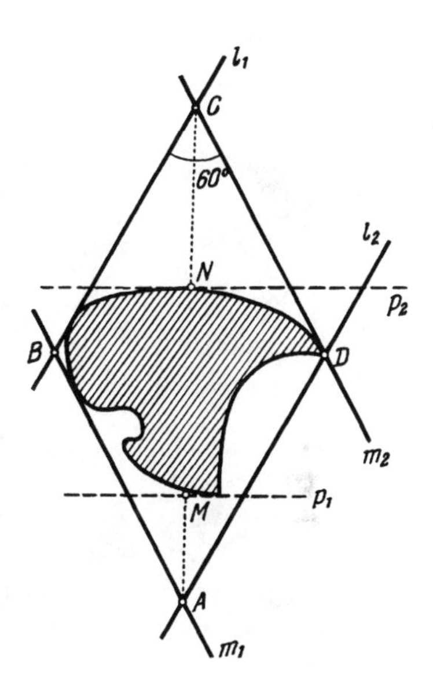

Fig. 1.15

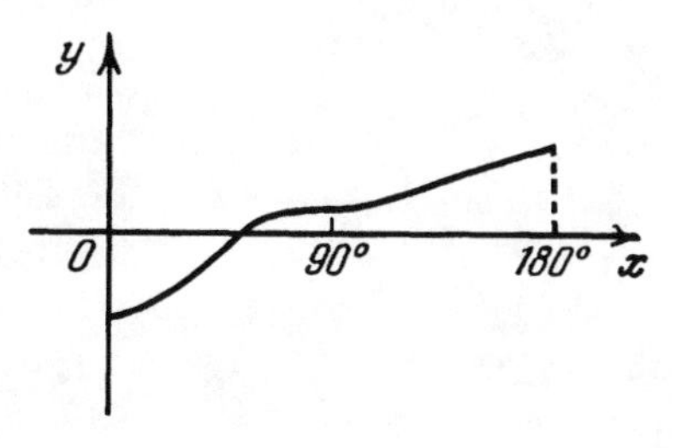

Fig. 1.16

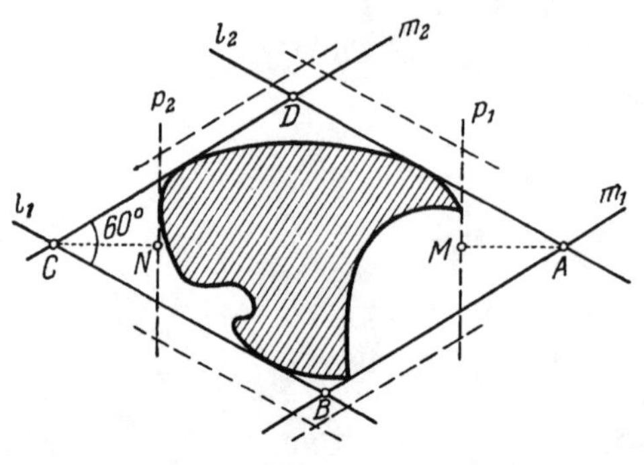

Fig. 1.17

(fig. 1.17). From the equality $AM = CN$ it follows that the hexagon formed by the lines l_1, l_2, m_1, m_2, p_1, and p_2 is symmetric with respect to its center. Every angle of this hexagon is 120°, and the distance between opposite sides does not exceed d. If the distance between p_1 and p_2 is *less than d*, we move these lines apart, moving them equal distances, just far enough that the distance between them will be d. We do the same for the lines l_1 and l_2, and then for m_1 and m_2, obtaining as a result a centrally symmetric hexagon with interior angles of 120° and a distance of d between each pair of opposite sides (the dotted hexagon in fig. 1.17). It is now clear that all sides of this hexagon are equal, and thus that it is regular. Moreover, it contains the figure F, and thus satisfies the specifications of the lemma.

Proof of theorem 1.1. Let F be a figure of diameter d. According to the lemma, F is contained in a regular hexagon whose opposite sides are separated by a distance of d. We shall show that this regular hexagon can be cut into three pieces, each of which has diameter less than d. The same division will, of course, also cut the figure F into three pieces, each having diameter less than d. The required partition of the regular hexagon into three pieces is given in figure 1.18 (the points P, Q, and R, are midpoints of sides, and O is the center of the hexagon). In order to convince ourselves that the diameters of the pieces are less than d, it suffices to observe that the angle PQL is a right angle, and therefore $PQ < PL = d$.

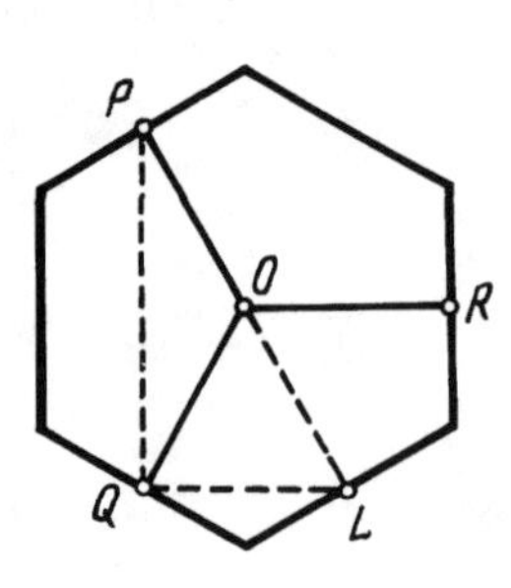

Fig. 1.18

The proof of theorem 1.1 is thus complete.

From the proof of theorem 1.1 it easily

follows that *any bounded plane figure of diameter d can be divided into three pieces, each having diameter less than or equal to* $(\sqrt{3}/2)d \approx 0.8660d$ (since it follows from $PL = d$ that $PQ = d\sqrt{3}/2$; see fig. 1.18). This upper bound on the diameters of the pieces is the best possible, for it is easy to see that *a circular disk of diameter d cannot be divided into three pieces, each having diameter less than* $d\sqrt{3}/2$. Indeed, the portion of the circumference covered by a piece with diameter less than $d\sqrt{3}/2$ must be contained in an arc of less than 120°, and therefore three such pieces cannot cover the whole circumference.

1.4. Convex Figures

Borsuk's theorem still does not give a complete solution to the problem of evaluating $a(F)$ for an arbitrary figure F of diameter d. It gives only an upper bound for $a(F)$: $a(F) \leq 3$. At the same time it is obvious that $a(F) \geq 2$ for any figure F. The problem which naturally arises is that of determining for which plane figures F the number $a(F)$ is 2 and for which it is 3. A solution to this problem will be given in sec. 1.7. Before arriving at this solution, however, we shall require some knowledge about *convex figures*, which we shall examine in this section and in the next two sections, generally with pictorial explanations rather than proofs.[1]

A figure is called *convex* if for each pair of its points A and B it contains the entire line segment AB connecting them (fig. 1.19). Thus, a triangle, a parallelogram, a trapezoid (all with their respective interior regions), a circular disk, a sector of a circular disk, and an ellipse (with its interior region) are all examples of convex figures (fig. 1.20). Figure 1.21 depicts some examples of nonconvex figures. The figures depicted in figure 1.20 are bounded; however, there are also *unbounded* convex figures (which "extend to infinity"): a half-plane, an angular region of less than 180°, and others (fig. 1.22).

The points of any figure may be divided into two classes: *interior* points and *boundary* points. The interior points are the points surrounded on all sides by points of the figure. Thus if A is an interior point

1. More detailed information about convex figures (and, in particular, proofs of the properties mentioned here) can be found in the following books: L. A. Lyusternik's *Convex Figures and Polyhedra* published in English by D. C. Heath and Company, New York, 1963 and (in paperback) 1966; and I. M. Yaglom's and V. G. Boltyanskii's *Convex Figures*, published in English by Holt, Rinehart, and Winston, New York, 1961. The article "Convex Figures and Solids," in volume 5 of the [Russian] *Encyclopedia of Elementary Mathematics* (pp. 181–269) is also devoted to this subject.

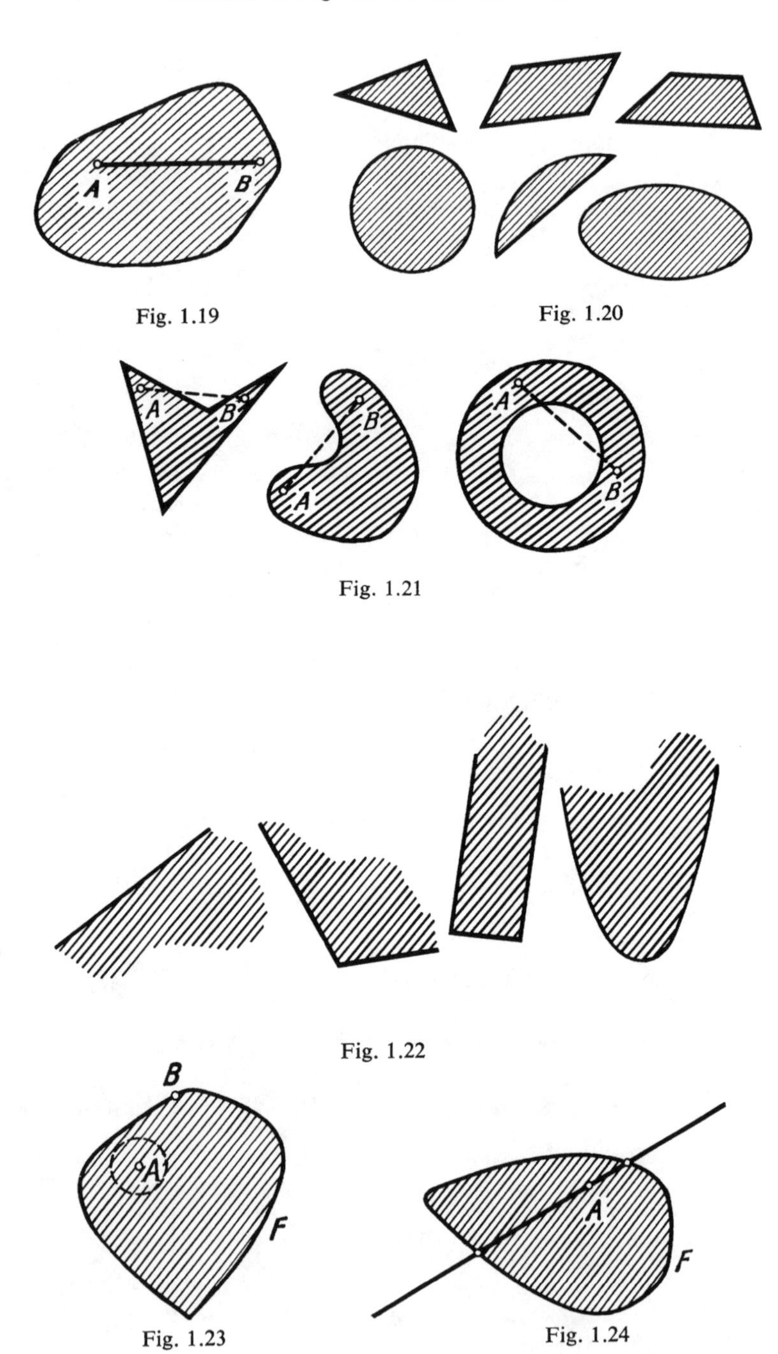

Fig. 1.19

Fig. 1.20

Fig. 1.21

Fig. 1.22

Fig. 1.23

Fig. 1.24

of F, then a disk of some (possibly very small) radius with center at A is entirely contained in F (fig. 1.23). A boundary point of F, on the other hand, is approached arbitrarily closely both by points of F and by points not belonging to F (for example, the point B in fig. 1.23). All of the boundary points together form a path, which is called the *boundary* of F. If a convex figure is bounded, then its boundary is a closed curve (figs. 1.19 and 1.20).

It is important for what follows to note that *any straight line passing through an interior point of a convex bounded figure F intersects the boundary of this figure in exactly two points* (fig. 1.24); the segment connecting these two points is contained entirely in F, and the remainder of the line lies outside F.

Let B be a boundary point of a convex figure F. The collection of all rays with endpoint B that pass through other points of F fills either a half-plane (fig. 1.25*a*) or an angular region of less than 180° (fig. 1.25*b*).

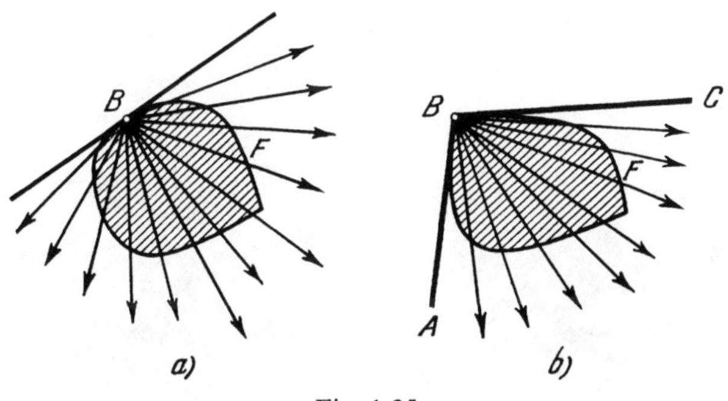

Fig. 1.25

In the former case, the line bounding the half-plane is a supporting line of F. Any other line through B will cut the figure into two parts (fig. 1.26)—that is, will not be supporting. In this case, then, there is a unique supporting line to the figure F passing through the point B; it is called the *tangent* to F at B. In the second case (fig. 1.25*b*), all of F lies within the angle ABC which is less than 180°, and therefore there are infinitely many supporting lines to F passing through B (fig. 1.27). In particular, the lines BA and BC are supporting. The rays BA and BC (heavily shaded in fig. 1.27) are called *semitangents* to F at B.

In either case, we see that *through each boundary point B of a convex figure F there passes at least one supporting line to F.* If only one supporting line of F passes through B (fig. 1.25*a*), then B is called an *ordinary*

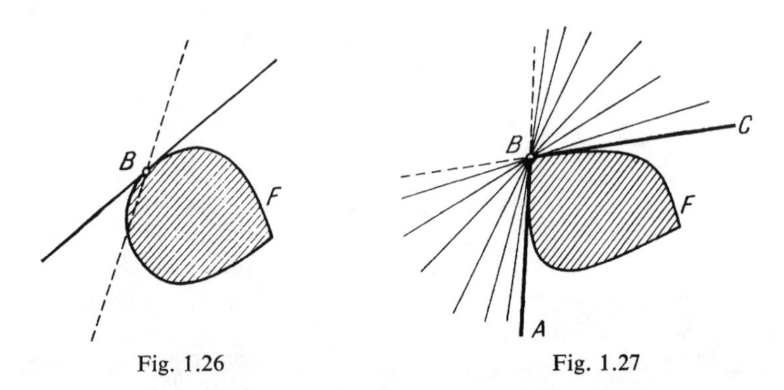

Fig. 1.26

Fig. 1.27

boundary point of F. If infinitely many supporting lines to F pass through B, B is called a *corner* point (fig. 1.25*b*).

Now let F_1 and F_2 be two convex figures. The *intersection* $F_1 \cap F_2$ is also a convex figure (fig. 1.28). Figure 1.29 depicts two convex figures: a circular disk F_1 and an angular region F_2 with vertex at the center of

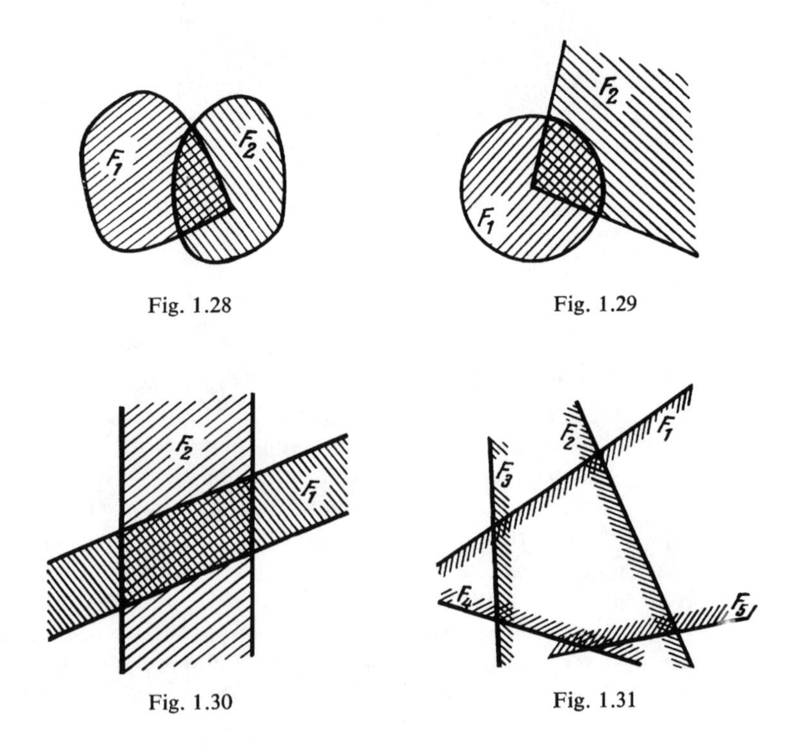

Fig. 1.28

Fig. 1.29

Fig. 1.30

Fig. 1.31

the disk; the intersection of these two figures is a sector of the disk. In figure 1.30, both of the convex figures F_1 and F_2 are unbounded (each is an infinitely extending strip); their intersection is a parallelogram. The foregoing applies not only to pairs of figures but to larger numbers of figures as well: *the intersection of any number of convex figures (even an infinite number) is a convex figure* (see remark 8). Figure 1.31 shows that a convex polygon is the intersection of a finite number of half-planes. A circular disk (fig. 1.32) is also an intersection of half-planes, but of an infinite number of them. In general, any convex figure can be represented as the intersection of an infinite number of half-planes.

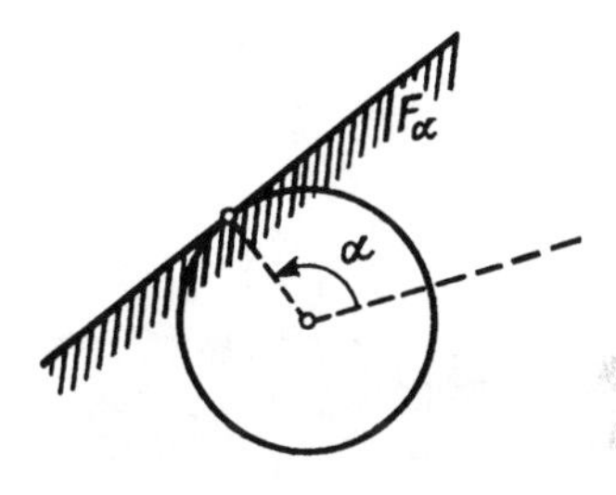

Fig. 1.32

Let F be a convex figure and A a point outside F. Then there exists a line *separating* A from F; that is, a line not containing A and not intersecting F placed in such a way that A and F are on opposite sides (fig. 1.33). This property of convex figures is characteristic: *if each point outside a given figure can be separated from the figure by a line, then the figure is convex.* In other words, if a figure F is not convex, then there is some point outside F which cannot be separated from F by any line (fig. 1.34).

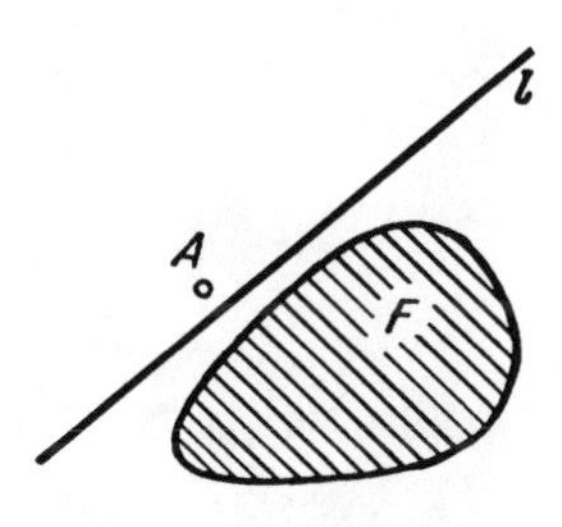

Fig. 1.33

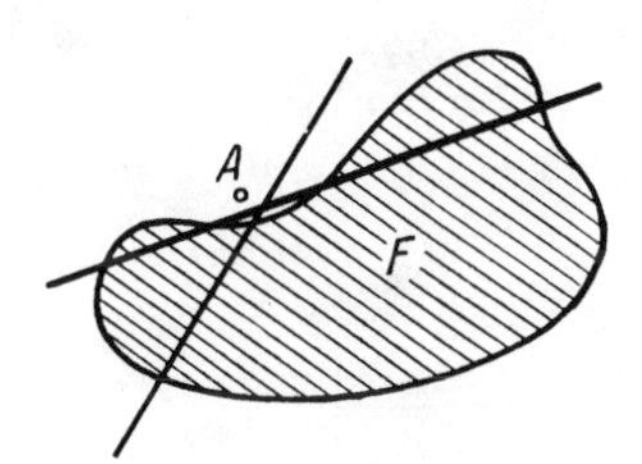

Fig. 1.34

In conclusion, let us note that for any bounded figure F of diameter d, there exists a "smallest" convex figure containing F (here denoted by $\tilde{F}$); this convex figure (fig. 1.35), called the *convex hull* of F, has the same diameter d. It may be that F is *disconnected*—that is, consists of two or more separated pieces—but even in this case, it is possible to define the convex hull of F. For example, figure 1.36 pictures the

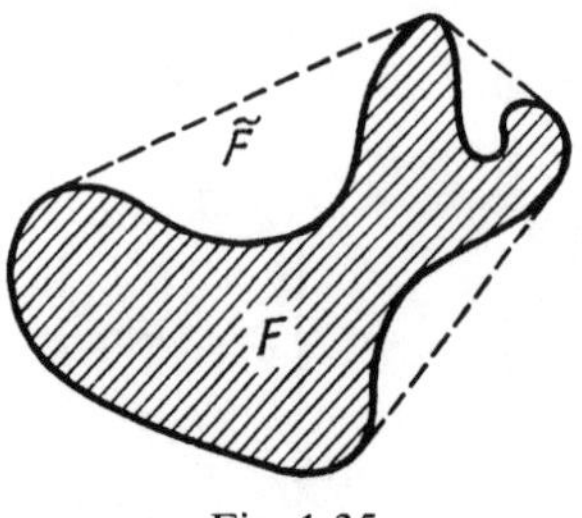

Fig. 1.35

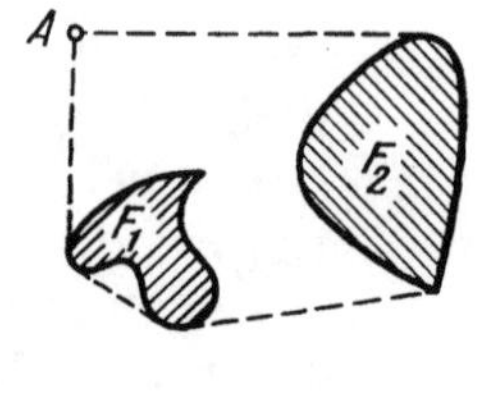

Fig. 1.36

convex hull of a figure which consists of two separate pieces F_1 and F_2 and a point A.

The convex hull of a plane figure can be visualized in the following way: if we wrap a rubber band around the figure, the curve formed by that rubber band will be the boundary of the convex hull. But this is only a pictorial explanation; the convex hull of F is formally defined as the intersection of all convex figures containing F. Indeed, as we have already noted, this intersection will be a convex figure. It is also clear that this intersection contains F and is the *smallest* convex figure having this property.

1.5. Figures of Constant Width

Let F be a bounded convex figure, and let l be a line. Draw two supporting lines to F parallel to l. The distance h between these two supporting lines is called the *width of F in the direction l.* Examining figure 1.37, one can easily see that the altitude of an equilateral triangle is its least width, and its side-length is its greatest width. A circular disk

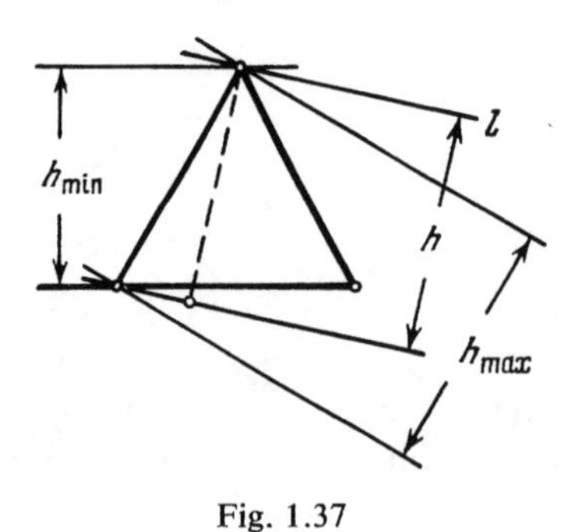

Fig. 1.37

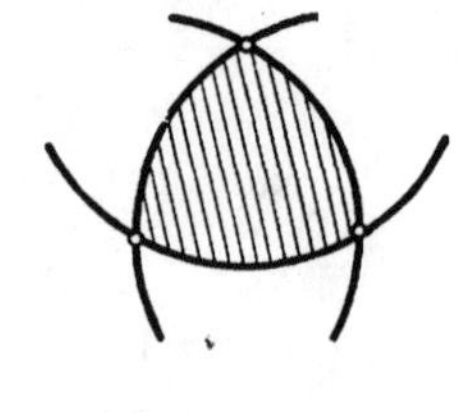
Fig. 1.38

has the same width (its diameter) in every direction. It may appear at first glance that a circular disk is the only kind of convex figure with this property. But this is not so: infinitely many shapes are possible for *figures of constant width*—that is, for convex figures having the same width in every direction. After the disk, the simplest example of such a figure is *Revleaux's triangle*, shown in figure 1.38. Revleaux's triangle is the intersection of the three disks of radius h whose centers are the vertices of an equilateral triangle with side h.

In general, if F is a regular polygon with an odd number of vertices and h is the length of the longest of the diagonals of F, then, by connecting each pair of adjacent vertices with a circular arc of radius h centered at the opposite vertex, we obtain a figure of constant width h (fig. 1.39). This construction also works for an irregular polygon having diameter h and an odd number of sides, if there are two diagonals of length h from each vertex (fig. 1.40).

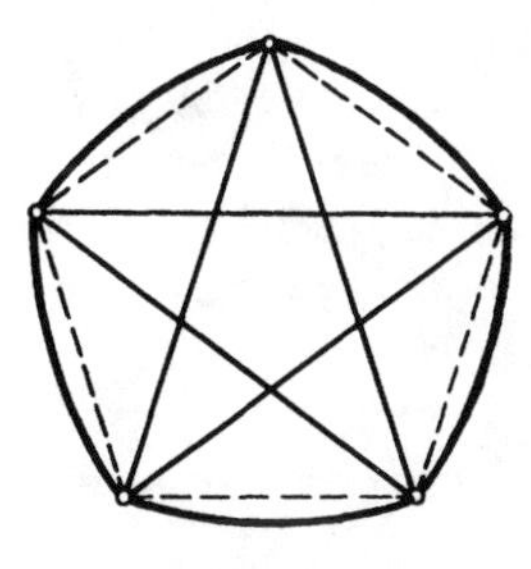

Fig. 1.39

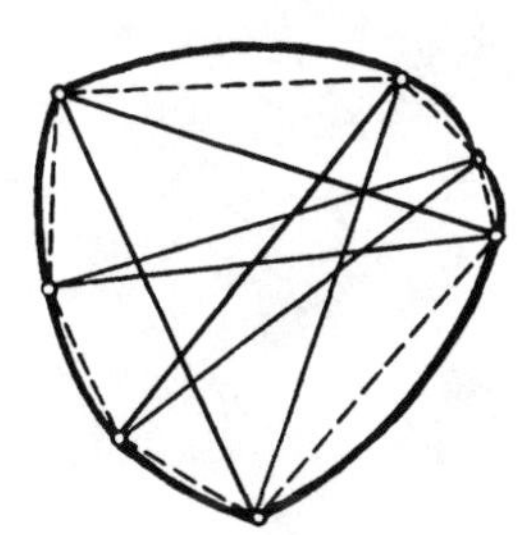

Fig. 1.40

Figures of constant width possess a number of interesting properties; we shall point out only a few elementary ones.[2]

First, we note that the diameter of a figure of constant width is equal to its width: $d = h$. Through each boundary point of such a figure there passes at least one *diameter* (that is, a chord of length d). It follows that *for any figure F of constant width*, $a(F) = 3$. In fact, one cannot even divide the *boundary* of a figure of constant width into two pieces of smaller diameter.[3] This is proven in the same way as in the case of a circle (p. 3 and remark 3).

2. Proofs of these properties can be found in the book by I. M. Yaglom and V. G. Boltyanskii mentioned on page 7.

3. The diameter of any bounded figure coincides with the diameter of its boundary (compare remark 2).—Trans.

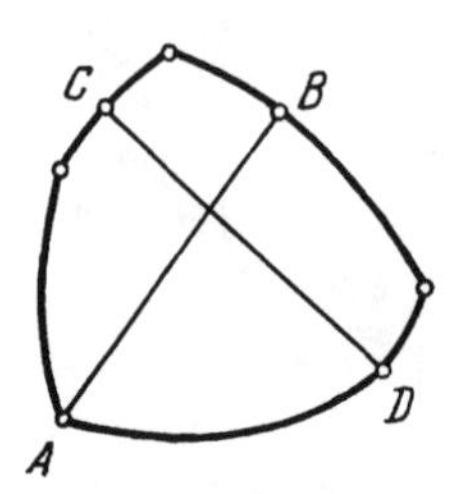

Fig. 1.41

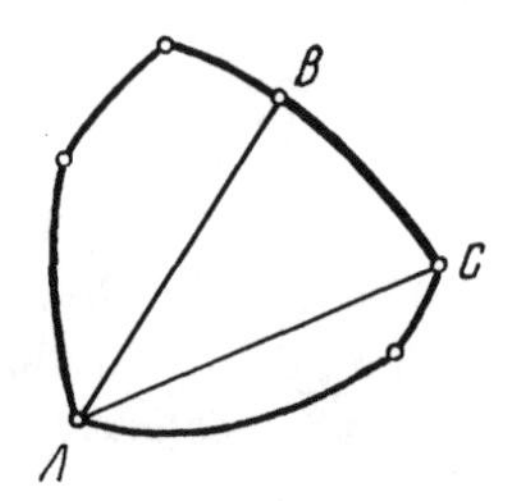

Fig. 1.42

Any two diameters of a figure of constant width intersect (either inside the figure or on its boundary; figs. 1.41 and 1.42). Furthermore, if two diameters AB and AC intersect at the boundary point A, then the entire circular arc BC with radius d and center A lies on the boundary of the figure (fig. 1.42), since each segment connecting A to a point between B and C on the boundary determines a width of the figure in some direction.

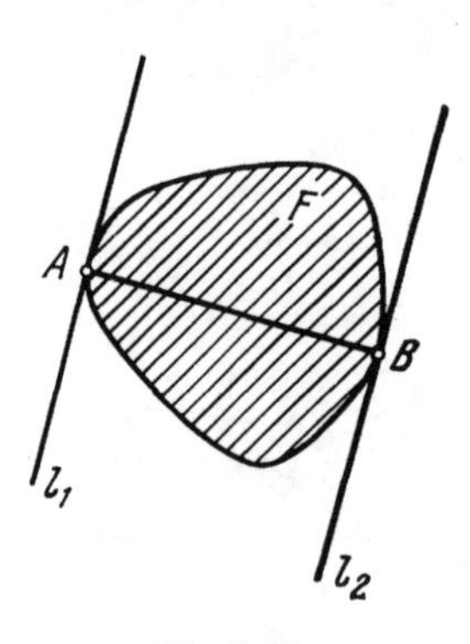

Fig. 1.43

Finally, we note that if F is a figure of constant width and AB is one of its diameters, then the lines l_1 and l_2 which pass through A and B, respectively, and which are perpendicular to the segment AB, are supporting lines of F (fig. 1.43).[4]

1.6. Embedding in a Figure of Constant Width

Returning to figure 1.39, let us denote our regular polygon by M, and the figure of constant width which contains it by F. Then the polygon M of diameter d is contained in the figure F of constant width d.

The construction mentioned in the last section does not work for regular polygons with an even number of sides. However, even in this case it is true that a regular polygon of diameter d can be embedded in a figure of constant width d (fig. 1.44). It is interesting to note that for a regular polygon with an *odd* number of sides, there is a *unique* figure of constant width which contains it and has the same diameter.

4. This property (along with its higher dimensional analogues with supporting planes or hyperplanes) holds not only for figures of constant width, but for all bounded figures.—Trans.

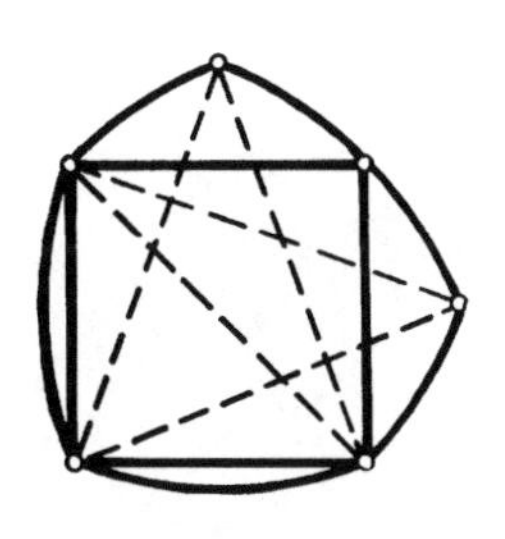

Fig. 1.44

But for a regular polygon with an even number of sides, there may be many such figures of constant width. For example, a square of diameter d is contained not only in the figure shown in figure 1.44, but also in the circumscribed disk, which is manifestly a figure of constant width d.

The above comments on the possibility of embedding regular polygons in figures of constant width have a far-reaching generalization:

THEOREM 1.2. *Any figure of diameter d can be embedded in a figure of constant width d.*

In preparation for the proof of this theorem, we shall introduce one new concept and prove three lemmas.

Let F be a plane figure of diameter less than or equal to d. It is clearly possible to find a disk of *radius d* which completely contains F (for example, if A is an arbitrary point of F, then the disk with radius d and center A completely contains F). We shall call the intersection of *all* disks of radius d containing F the *d-extension* (or simply the *extension*) of F, and denote it by F^*. For example, if F is an equilateral triangle with side d, then F^* is Revleaux's triangle (fig. 1.45).

It is clear from the definition that the d-extension of a figure of diameter not exceeding d is a convex figure, as it is an intersection of convex figures.

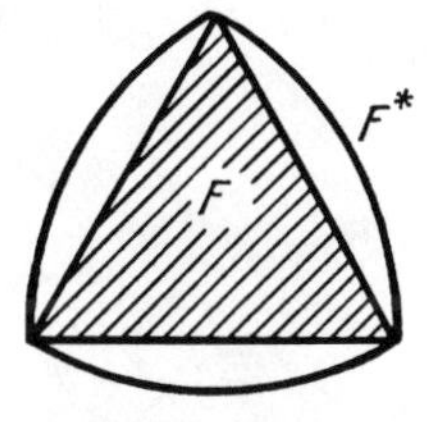

Fig. 1.45

LEMMA 1.2. *Let F be a figure of diameter d. Then its d-extension F* also has diameter d* (see remark 9).

Proof. Let A and B be arbitrary points of F^*. Take an arbitrary point M of F and consider the disk K_M of radius d centered at M (fig. 1.46). This disk completely contains F and thus is a member of the set of disks whose intersection is F^*. Therefore K_M also contains F^*. In particular, the point A belongs to K_M, and therefore

$$AM \leq d\,.$$

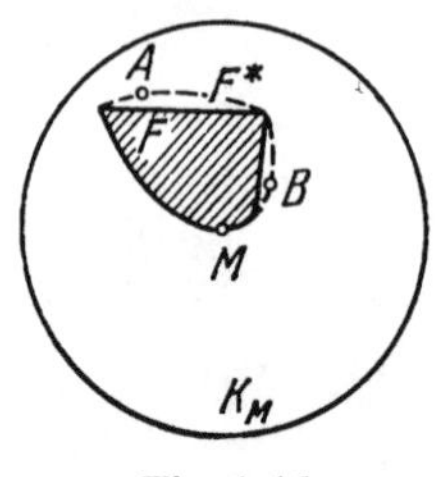

Fig. 1.46

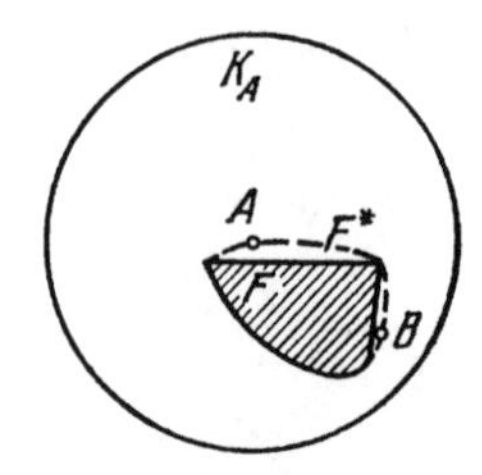

Fig. 1.47

Thus, since M was arbitrary, $AM \leq d$ for any point M of F, and the disk K_A with radius d and center A contains all of F (fig. 1.47). K_A is therefore one of the disks whose intersection is F^*, and it follows that F^* is entirely contained in K_A. In particular, the point B is in K_A, so that $AB \leq d$.

Thus, the distance between any two points A and B of the figure F^* does not exceed d; that is, the diameter of F^* is no greater than d. But the diameter of F^* is also no less than d, for F^* completely contains F, a figure of diameter d.

LEMMA 1.3. *Let F be a figure of diameter d with d-extension F^*, let M be an arbitrary boundary point of F^*, and let l be a supporting line to F^* passing through M. Let K_l be the disk of radius d which is tangent to the line l at M and which lies on the same side of l as F does. Then K_l completely contains F^** (fig. 1.48).

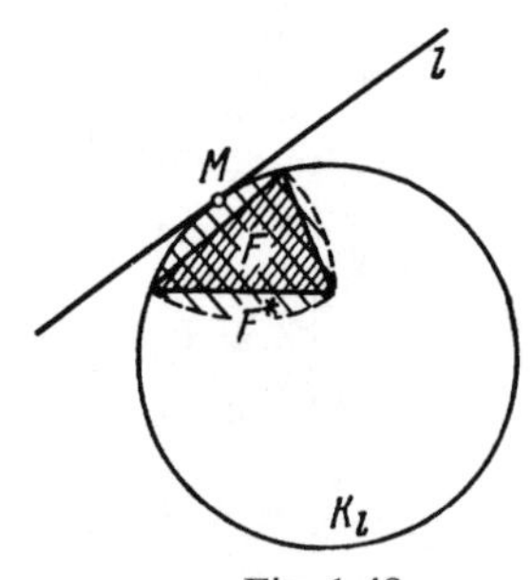

Fig. 1.48

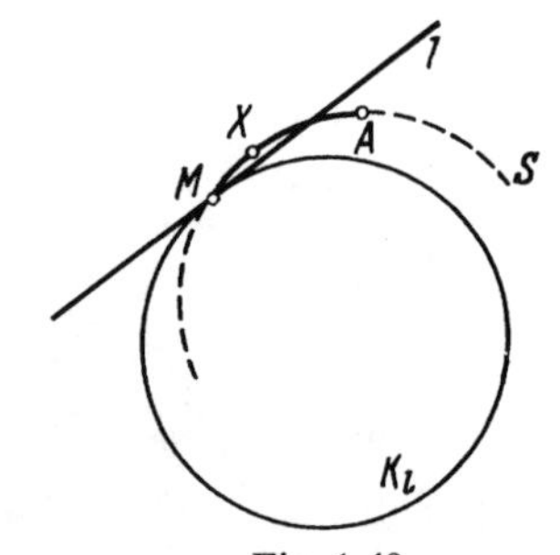

Fig. 1.49

Proof. Suppose K_l does not contain F^*; that is, suppose some point A of F^* lies outside K_l. Draw a circle S of radius d through the points M and A, with center on the same side of l as F^* (fig. 1.49). Since S is distinct from the circumference of the disk K_l, S is *not tangent* to l at M, and so S crosses l. Consequently, the minor arc AM of S crosses l (and not the major arc, since the center of S lies on the same side of l as F^*, and thus on the same side as A, a point of F^*). That is, there is some point X on the minor arc AM which lies on the opposite side of l from F^*. Now let K be an arbitrary disk of radius d containing F. Then K contains F^*, since K is a member of the family of disks whose intersection is F^*, and therefore the points A and M belong to K. It follows that K contains all of the arc AM (since AM is a minor arc of the circle S, which has the same radius as the disk K); in particular, K contains the point X. Since K was arbitrary, any disk of radius d containing F also contains X, and therefore X belongs to F^*, contradicting the assumption that l is a supporting line of F^*. Thus our assumption that some point A of F^* lies outside K_l is false, and the disk K_l completely contains F^*.

LEMMA 1.4. *Let F be a convex figure of diameter d. If F does not have constant width, then there exists a bounded convex figure H of diameter d properly containing F (and consequently having a larger area)* see remark 10).

Proof. Consider the extension F^* of F. If F does not coincide with F^*, then the figure

$$H = F^*$$

is the desired figure: it contains F, has diameter d, and obviously has area greater than that of F.

Now suppose that F coincides with F^*. Since F does not have constant width d, F must have two parallel supporting lines l' and l'' which are separated by some distance less than d. Let M be a point common to l'' and F, and let K denote the disk of radius d tangent to l'' at M which lies on the same side of l'' as does F (fig. 1.50).

Let A denote the center of the disk K. A does not belong to F, because $AM = d$, and the distance between l' and l'' is less than d. By lemma 1.3, K completely contains F and F^* (which in this case happen to coincide). Consequently, $AP \leq d$ for every point P of F. In other words,

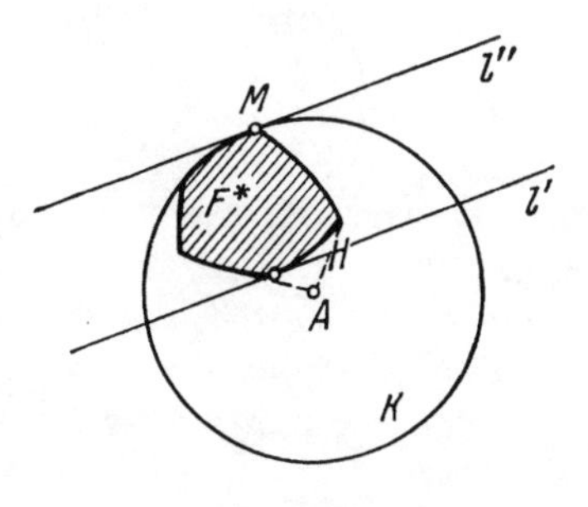

Fig. 1.50

the figure F' consisting of F and the point A has diameter d. The convex hull H of F' (fig. 1.50) also had diameter d. Since the area of H is obviously greater than that of F (we recall that the point A does not belong to F), H is the desired figure.

Proof of theorem 1.2. Let F be a figure of diameter d. If its extension F^* is a figure of constant width, the conclusion of the theorem is true.

Now suppose that F^* is not a figure of constant width. Let us consider all figures of diameter d which contain F^*. Since every figure of diameter d is entirely contained in some disk of radius d, the area of any such figure cannot exceed πd^2. Let k_0 be the greatest integer such that there exists a convex figure of diameter d containing F^* and having area greater than or equal to k_0. Choose one of these figures and call it H_0. Then H_0 has diameter d, and it is clear that the area of any convex figure of diameter d which contains H_0 exceeds the area of H_0 by less than 1 (by our choice of k_0).

Now let k_1 denote the largest integer such that there exists a convex figure of diameter d which contains H_0 and has area greater than or equal to $k_0 + \frac{1}{10}k_1$. Select one such figure and call it H_1. Then H_1 contains H_0, has diameter d, and has an area which comes within $\frac{1}{10}$ of the area of any convex figure of diameter d containing it.

In the same way, we construct a convex figure H_2 of diameter d containing H_1 in such a way that any convex figure of diameter d containing H_2 exceeds H_2 in area by less than $1/100$. Then we go through a similar construction for $1/1000$, and so on.

This procedure has two possible results: either at some step we obtain a figure H_n which has constant width d (in which case the construction is finished), or none of the figures H_n has constant width, in which case lemma 1.4 provides us with an infinite sequence of convex figures $H_0, H_1, \ldots, H_n, \ldots$, each of which is contained in the next and each of which has diameter d. In the second case, let H denote the union of all the figures $H_0, H_1, \ldots, H_n, \ldots$ (see remark 11).

The set H is convex, for if A and B are points of H, A belongs to some H_n and B to some H_m. Without loss of generality, we may assume that $n \geq m$. Then H_m is entirely contained in H_n and consequently both A and B belong to H_n. Since H_n is convex, it must contain the entire segment AB as well. But then the segment AB is entirely contained in H; thus H is convex.

Furthermore, it is easy to see that the diameter of H is d. For if A and B are two points of H, then, as before, both A and B belong to some common H_n. Since the diameter of H_n is d, $AB \leq d$. Yet H_0 (and thus

H) contains two points C and D for which $CD = d$; thus, by our definition of diameter, H has diameter d.

Finally, we claim that H has constant width d; for if not, then by lemma 1.4 there exists a convex figure H' of diameter d which contains H and has area greater than that of H. Let n be a natural number such that the difference between the areas of H' and H is greater than or equal to $1/10^n$. Then the difference between the areas of H' and H_n, which is larger, must be greater than $1/10^n$. But this contradicts our choice of H_n. Thus H is of constant width d, and the proof of the theorem is complete.

1.7. For Which Figures Is $a(F) = 3$?

As we have already noted, in some instances a figure of diameter d can be enlarged to a figure of constant width d in only one way (as in the case of a regular polygon with an odd number of sides). In other instances, this completion is not unique. As one more example of a figure that admits of more than one completion to a figure of constant width, consider a disk minus a segment of that disk whose arc is less than a semicircle. The diameter d of this figure F is clearly equal to the diameter of the original disk. Therefore the disk is one of the figures of constant width d containing F. Another completion to a figure of constant width is shown in figure 1.51 (in which $AN = NB = AC = BD = d$, and the arcs AB, DN, and CN are circular arcs of radius d).

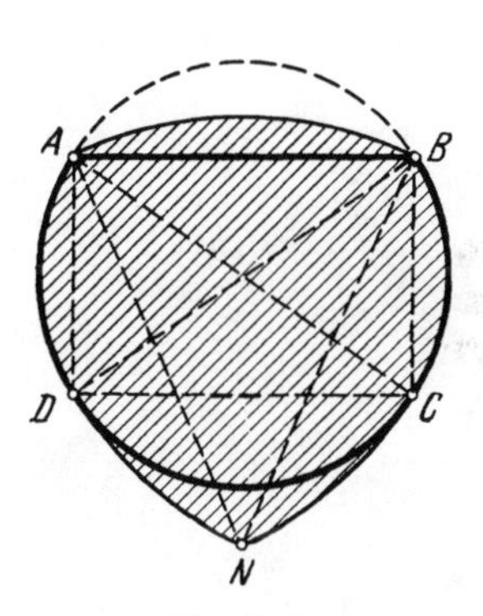

Fig. 1.51

We now return to the problem formulated at the beginning of sec. 1.4: to find all figures F for which $a(F) = 3$. The following solution was discovered in 1969 by V. G. Boltyanskii.

THEOREM 1.3. *Suppose F is a plane figure of diameter d. Then $a(F) = 3$ if and only if F has a unique completion to a figure of constant width d.*

In particular, $a(F) = 3$ for any regular polygon F with an odd number of sides, and $a(F) = 2$ for any regular polygon F with an even number of sides (the latter statement, however, is immediately evident; see fig. 1.52). Moreover, $a(F) = 2$ for a disk from which a segment has been removed (cf. fig. 1.51); this, too, can be shown without the help of the theorem (fig. 1.53).

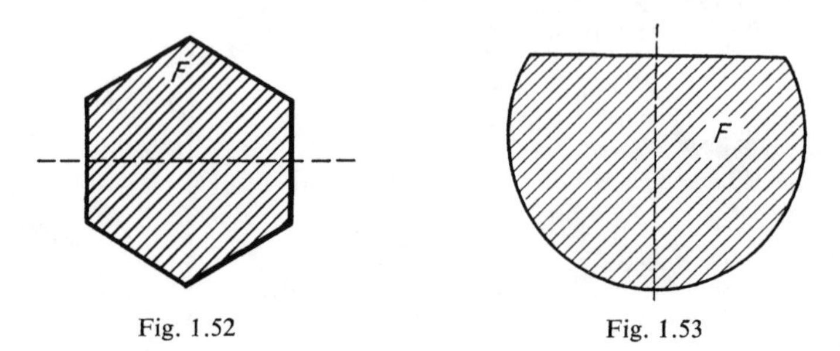

Fig. 1.52 Fig. 1.53

Before turning to the proof of theorem 1.3, we shall prove several auxiliary propositions.

LEMMA 1.5. *Let F be a convex figure of diameter d. If F is contained in some disk of radius d whose center does not belong to F*, then the completion of F to a figure of constant width d is not unique.*

Proof. Let K be a disk of radius d containing F and having its center A outside F^*. Since F^*, the intersection of all disks of radius d containing F, does not contain A, there is some disk K' of radius d which contains the figure F but not the point A. Denote the center of K' by B (fig. 1.54). Then the segment AB has length greater than d, since A lies outside the disk K'.

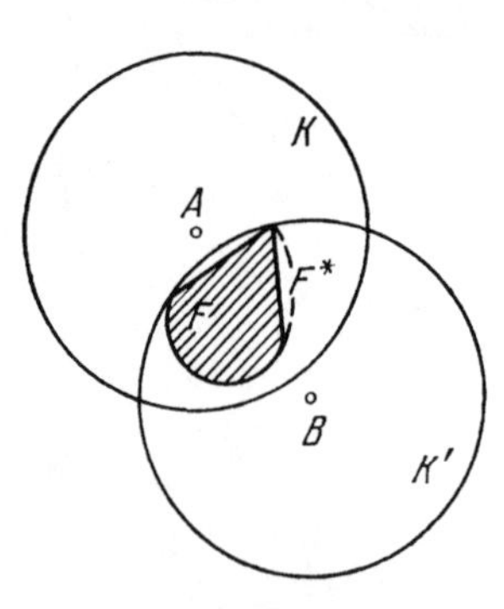

Fig. 1.54

Now adjoining the point A to the figure F, we obtain the disconnected figure F', the diameter of which is easily seen to be d. (The distance from the "new" point A to any point of F cannot exceed d, since F is contained in the disk K.) In the same way, adjoining the point B to F, we obtain the disconnected figure F'', the diameter of which is d. Applying theorem 1.2, we may choose some figure Φ' of constant width d containing F', and some figure Φ'' of constant width d containing F''. The figures Φ' and Φ'' clearly do not coincide (since $AB > d$, and therefore no figure of constant width d can contain both A and B). At the same time, each of the figures Φ' and Φ'' contains F. Thus F can be completed to a figure of constant width d in at least two ways.

Lemma 1.6. *Let F be a figure of diameter d, and let Φ be a figure of constant width d containing F. Then the figure F^* is entirely contained in Φ.*

Proof. Let A be an arbitrary point outside Φ. Draw a line l separating A from Φ. Let l' and l'' denote supporting lines of Φ parallel to l, and let B and C be the points where these lines touch Φ (fig. 1.55). Then the segment BC is perpendicular to l' and l'' (otherwise the width of Φ in the direction perpendicular to BC would be greater than that in the direction l) and has length d. Denote by K_C the disk of radius d centered at C. This disk completely contains Φ, and thus contains F. K_C is therefore one of those disks whose intersection defines F^*, and consequently F^* is entirely contained in K_C. Since A does not belong to K_C, A cannot belong to F^*. Thus, if a point A lies outside Φ, it also lies outside F^*. In other words, F^* is contained in Φ.

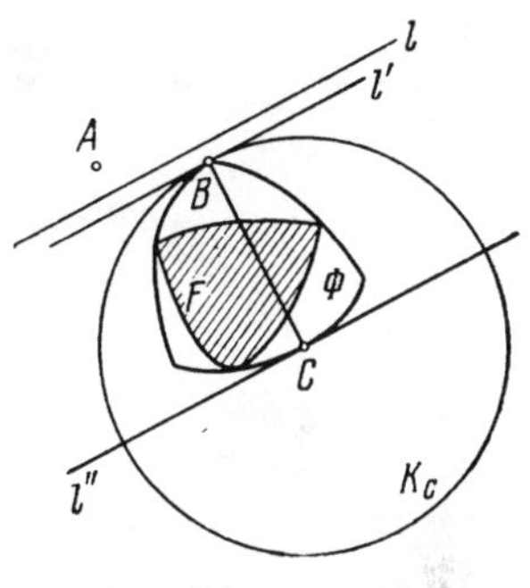

Fig. 1.55

Lemma 1.7. *A figure F of diameter d has a unique completion to a figure of constant width d if and only if its extension F^* has constant width d.*

Proof. Suppose F^* has constant width d, and let H be an arbitrary figure of constant width d containing F. Then, by lemma 1.6, H must contain F^*. Since F^* is already a figure of constant width d, H coincides with F^*. Thus the completion of F to a figure of constant width d is unique.

Conversely, suppose that F^* does not have constant width d. Then F^* has a pair of parallel supporting lines l and l' separated by a distance of less than d (fig. 1.56). Let M be the point where l touches F^*, and denote by K_l the disk of radius d tangent to l at M and located on the same side of l as F^*. By lemma 1.3, K_l contains F^*. Furthermore, it is clear that the center of K_l, which we shall call A, does not belong to F^* (for F^* lies

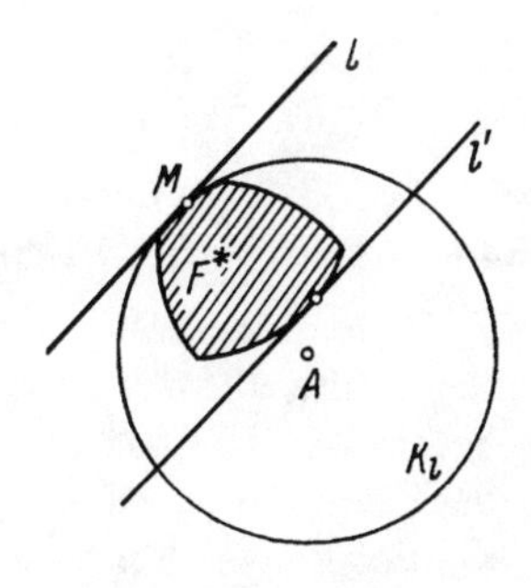

Fig. 1.56

in the strip between l and l', and A lies outside this strip). But then, by lemma 1.5, F has more than one completion to a figure of constant width d.

LEMMA 1.8. *Under the conditions of lemma 1.3, if the point M does not belong to F, then F contains two points A and B on the circumference of the disk K_l such that the minor arc AB of this circumference contains M.* (fig. 1.57).

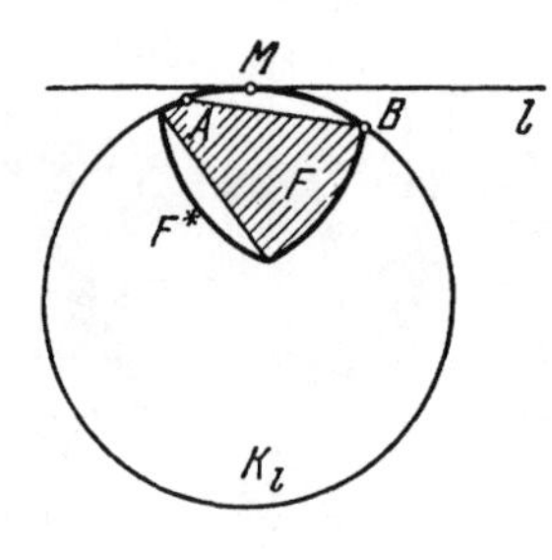

Fig. 1.57

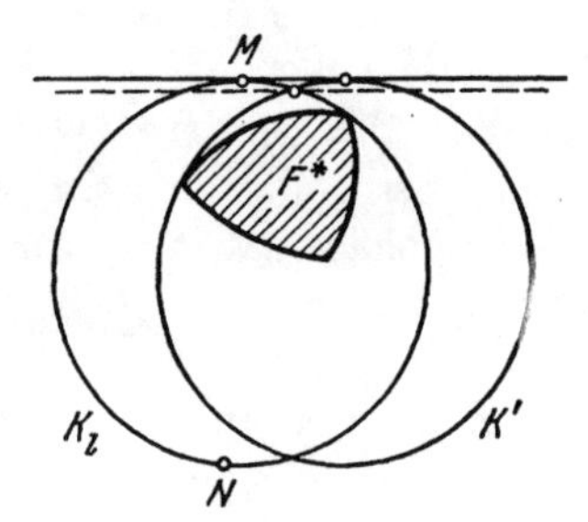

Fig. 1.58

Proof. Denote by N the point of K_l diametrically opposite M. For the sake of simplicity, let us agree to consider the line l "horizontal" and the disk K_l as lying "below" l (fig. 1.57). If the left semicircle determined by M and N did not contain any point of F, then K_l could be moved to the right, and the resulting disk K' would still contain F (and thus F^*). But then F^* would be contained in the intersection of K_l and K', and the line l could not be a supporting line to F^* (fig. 1.58). Thus, the left semicircle must contain at least one point A of F (fig. 1.57). Similarly, the right semicircle must contain some point B of F. Furthermore, the fact that A, B, and M belong to F^* implies that $AM \leq d$ and $BM \leq d$. Therefore, of the two arcs determined by A and B on the circumference of K_l, the one containing M is at most a semicircle (it does not even exceed 60°, since $AB \leq d$).

LEMMA 1.9. *If F is a figure of diameter less than d, then its d-extension F^* also has diameter less than d.*

Proof. Denote the diameter of F by d'; we then have $d' < d$. Further, let us denote the d'-extension of F by F', and the d-extension of F by F^*, as before. If a point M does not belong to F', then there is a disk K' of radius d' containing all of F but not containing M. Let us denote by A the point of K' which is *closest* to M (fig. 1.59) and construct the disk K of radius d which contains K' and is internally tangent to K' at A. K

clearly does not contain M, but contains all of F, and it follows that M does not belong to F^*. Therefore, if a point M does not belong to F', it does not belong to F^* either; thus F^* is entirely contained in F'. But F', as the d'-extension of the figure F of diameter d', must have diameter d' (by lemma 1.2). Hence the diameter of F^* is at most d', which is less than d.

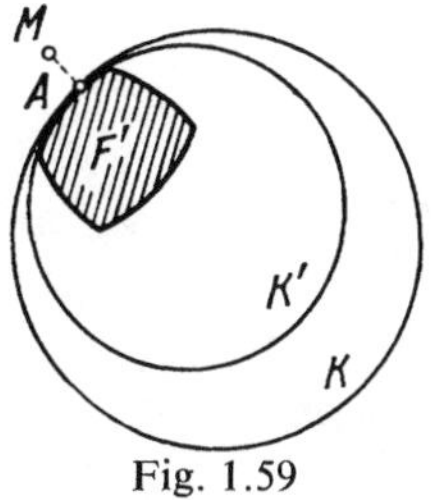

Fig. 1.59

Proof of theorem 1.3. Let F be a figure of diameter d having more than one completion to a figure of constant width d. We shall prove that $a(F) = 2$. For if Φ_1 and Φ_2 are two distinct figures of constant width d containing F, there must be some boundary point A of Φ_1 which lies in the interior of Φ_2 (for the two figures intersect, and neither can be properly contained in the other). Through this point A we may draw a supporting line l_1 to Φ_1. Let l_2 be another supporting line to Φ_1, parallel to l_1, and let B be the point at which this supporting line meets Φ_1 (fig. 1.60). Then the line AB is perpendicular to l_1 and l_2. We shall prove that the line AB cuts F into two parts, each of which has diameter less than d.

For suppose, on the contrary, that one of these parts has diameter d; that is, that on one side of the line AB there exist points C and D of F separated by a distance of d. Then the segments AB and CD are *diameters* of the figure Φ_1 (recall that F is contained in Φ_1 and that $AB = d$, since Φ_1 is a figure of constant width d); as Φ_1 has constant width, the segments AB and CD must therefore intersect, either at a point interior to both of them or at a common endpoint. The former case is impossible, since C and D are on the same side of the line AB. Thus the segments AB and CD must have a common endpoint. But B lies outside Φ_2 (as A lies in the interior of Φ_2 and $AB = d$), and thus outside F, whereas C and D belong to F. Hence B cannot be the common endpoint of the segments AB and CD. Finally, the distance between the interior point A of Φ_2 and any other point of Φ_2 must be *less* than d, and therefore A cannot coincide with an endpoint of CD, a segment of length d contained in Φ_2.

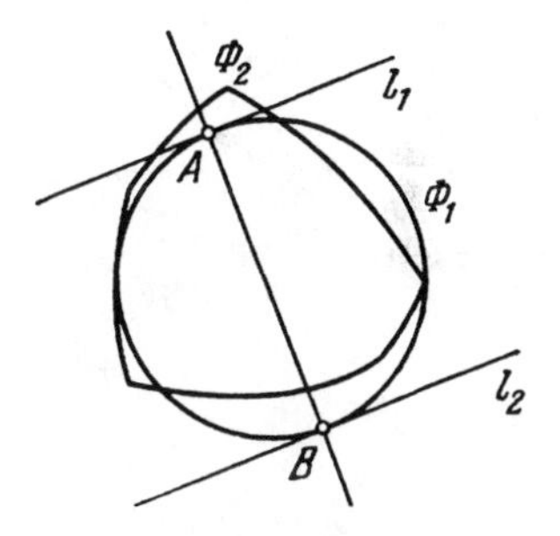

Fig. 1.60

The contradiction thus obtained shows that each of the pieces into which the line AB divides F has diameter less than d, and thus that $a(F) = 2$.

Now suppose, conversely, that F, a figure of diameter d, has a unique completion to a figure of constant width d. We shall show that, in this case, $a(F) = 3$. For suppose, on the contrary, that $a(F) = 2$; that is, that F can be represented as the union of two figures Q_1 and Q_2, each of which has diameter less than d. Since F has a unique completion to a figure of constant width d, F^* has constant width d (lemma 1.7). Denote the d-extensions of Q_1 and Q_2 by $Q_1{}^*$ and $Q_2{}^*$, respectively. By lemma 1.9, both $Q_1{}^*$ and $Q_2{}^*$ have diameter less than d.

Let Γ denote the boundary of F^*, and let M be an arbitrary point of the curve Γ. If M belongs to F, then M is obviously contained in the union of the figures Q_1 and Q_2, and thus in the union of Q_1^* and Q_2^*. If M does not belong to F, draw a supporting line l to the figure F^* passing through M, and construct the disk K_l of radius d tangent to l at M which is on the same side of l as F^* (fig. 1.61). The center N of K_l belongs to the boundary Γ of F^* (because MN is perpendicular to l and $MN = d$, and therefore MN must be a diameter of F^*, a figure of constant width d). Since M does not belong to F, lemma 1.8 implies the existence of two points A and B of F, lying on the circumference of K_l, such that the minor arc AB of this circumference contains the point M. Thus, $AN = BN = d$, so that AN and BN are diameters of F^*, which has constant width d. It follows that N is a *corner* point of the curve Γ and that the entire minor arc AB belongs to Γ, so that M is a *regular* (noncorner) point of Γ. But it is then clear that the point N *must* belong to F (for otherwise, by interchanging the roles of M and N, we could similarly deduce that M is a corner point of Γ and that N is a regular point of Γ).

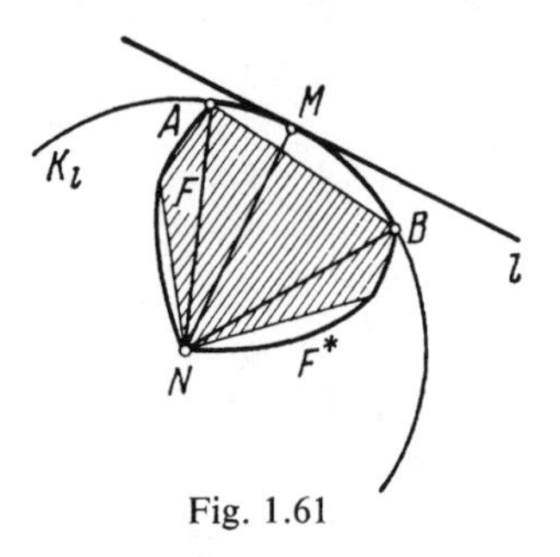

Fig. 1.61

Since N belongs to F, it must belong to at least one of the figures Q_1 or Q_2. Without loss of generality, say N belongs to Q_1. As $AN = BN = d$, and Q_1 has diameter less than d, the points A and B cannot belong to the set Q_1. But since both of these points belong to F, they must both lie in Q_2. The minor arc AB of radius d, which contains M, is entirely contained in any disk of radius d that contains A and B. In particular, this arc is contained in any disk of radius d that contains Q_2. Thus the intersection Q_2^* of all such disks contains the minor arc AB, and in particular the point M. Analogously, if N belongs to Q_2, then M belongs to Q_1^*. Thus, in any case, M belongs to the union of Q_1^* and Q_2^*.

Since M was arbitrary, we have proved that any point M of the curve Γ (regardless of whether it belongs to F) is contained in the union of the figures Q_1^* and Q_2^*. Thus, the entire curve Γ is contained in the union of Q_1^* and Q_2^*, so that Γ can be expressed as the union of two curves (one lying entirely in Q_1^*, the other in Q_2^*) of diameter less than d. But this is impossible, as Γ is the boundary of a figure of constant width (see page 13). Consequently, our assumption [that $a(F) = 2$] is false, and $a(F) = 3$ for a figure F of diameter d with a unique completion to a figure of constant width d.

The proof of theorem 1.3 is thus complete (see remark 12).

We observe in conclusion that theorem 1.3 can, by virtue of lemma 1.7, be formulated as follows:

Let F be a figure of diameter d. Then $a(F) = 3$ if and only if F^* has constant width d.

We note also the following curious fact:

For any bounded plane figure F, $a(F) = a(F^*)$.

Indeed, since F^* is its own d-extension, the following three statements are equivalent in the light of our new formulation of theorem 1.3: (*a*) $a(F^*) = 3$; (*b*) $(F^*)^* = F^*$ has constant width; (*c*) $a(F) = 3$.

2 Division of Figures in the Minkowski Plane

2.1. A Graphic Example

If a line segment LM is chosen as a unit of length, then the *length* of any line segment AB can be defined as the ratio AB/LM. The length of the segment AB depends only on its *magnitude* and not at all on its *direction* or location. Some problems, however, require a different definition of length, one in which a segment's length depends on its direction as well as on its magnitude. For such a definition, it is necessary to have a unit of length for each particular direction. A very interesting definition of this sort, which we shall consider in this section, was proposed at the end of the nineteenth century by the noted German mathematician Hermann Minkowski. Before giving this definition, we shall analyze a graphic example.

Picture a perfectly planned city M in which half of the streets are perfectly vertical and half of the streets are perfectly horizontal (fig. 2.1). Suppose someone wishes to travel from a point A to a point B in this city. What will be his notion of the "distance" between these two points?

On a map of the city one can, of course, draw the segment AB with a ruler and measure its length. But in this city such a concept of distance would be useless, since motion along the line segment AB might require the ability to pass through buildings. In this case, it is more useful to consider the distance between A and B to be the length of the broken line ACB shown in figure 2.2. Besides ACB there exist many other possible paths from A to B of the same length (fig. 2.3), but there are no shorter possible paths. If we introduce a system of coordinates for our map with axes on two perpendicular streets, the above discussion indicates that the practical distance between points $A(x_1, y_1)$ and

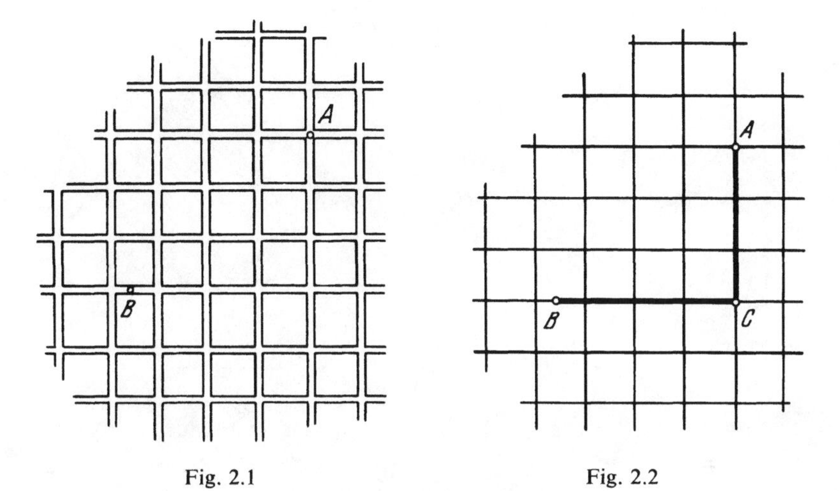

Fig. 2.1

Fig. 2.2

$B(x_2, y_2)$ of our city M—that is, the length of a "segment" (or shortest path) AB in the geometry of this city—is

$$d_m(A, B) = |x_2 - x_1| + |y_2 - y_1| \qquad (*)$$

(fig. 2.4). Now that we know how to find the "distance" between two points in the city M, we may consider the problem of finding the "unit disk" in our city, that is, the set of all points whose distances from the coordinate origin O is no more than 1. Since the point O has coordinates (0, 0), the formula (*) gives the distance from O to a point $C(x, y)$ as

$$d_m(O, C) = |x| + |y| .$$

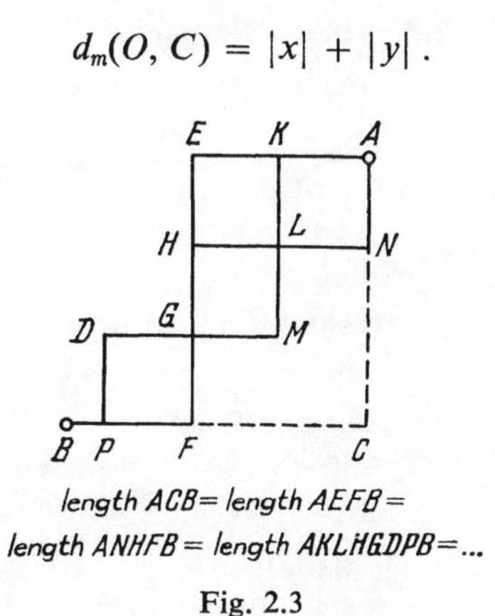

Fig. 2.3

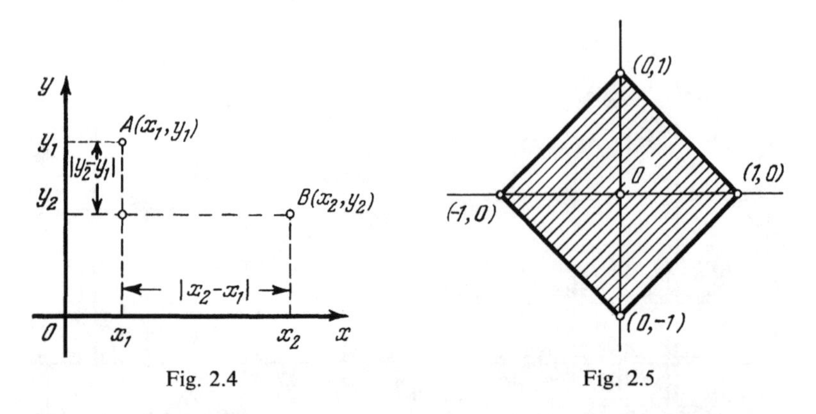

Fig. 2.4

Fig. 2.5

The "unit disk" in this city is therefore defined by the inequality

$$|x| + |y| \leq 1 .$$

It is now evident that the desired "unit disk" on a map of our city is actually a *square* (fig. 2.5). Using this unit disk, we can now find the distance between any two points in the manner indicated at the beginning of this section. Specifically, if we are given two arbitrary points A and B, we find on the boundary of the "unit disk" a point C such that OC is parallel to AB. The desired distance will then be the ratio $AB{:}OC$.

It is this idea—that of taking as the unit disk some convex *centrally symmetric* figure—that forms the basis of Minkowski's geometry in the plane.[1]

2.2. The Minkowski Plane

Suppose we are given a bounded plane convex figure G symmetric with respect to some interior point O (fig. 2.6).[2] Let Γ denote the curve bounding the figure G. The unit of length in any given direction l will be the segment OL of the ray with endpoint O parallel to l, where L is the point of intersection of this ray with Γ. The length of a segment AB will now be defined as the ratio $AB{:}OL$, where OL is a unit segment in the direction determined by the vector $\mathbf{AB}$. (If A coincides with B, we

1. For other methods of measuring distances (and for the most general mathematical concept of "distance"), the reader is referred to Yu. A. Shreider's *What Is Distance?*, published in English by The University of Chicago Press, 1974.

2. A figure F is said to be *symmetric* with respect to a point A if for each point B of F, the point C lying on the line AB "opposite" B (the same distance from A, but in the opposite direction) is also contained in F.

naturally consider the length of AB to be zero.) Henceforth we shall use the symbol $d_G(A, B)$ to denote the length of a segment AB in the system of measurement generated by the figure G. Obviously, $d_G(O, M) = 1$ if and only if M lies on the curve Γ. If M lies in the interior of G, then $d_G(O, M) < 1$; if M lies outside G, then $d_G(O, M) > 1$.

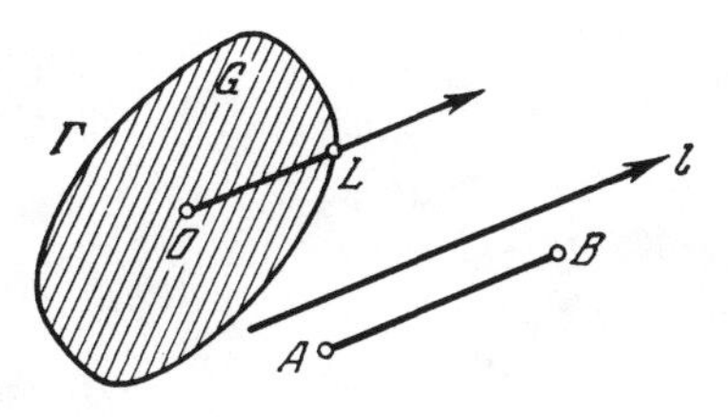

Fig. 2.6

Note that if G happens to be circular we have the usual definition of length, with the length of a segment depending only on its magnitude and not on its direction; and if G is a square (fig. 2.5), we arrive at the definition of length studied in the preceding section.

We shall now consider the basic properties of our new definition of length. As is clear from the definition,

$$d_G(A, B) \geq 0 .$$

Since O is an interior point of G, equality holds if and only if A and B coincide. Further, the symmetry of G with respect to its center implies that

$$d_G(A, B) = d_G(B, A) .$$

Finally, if AB and CD are parallel segments and if $AB/CD = k$, then

$$\frac{d_G(A, B)}{d_G(C, D)} = k .$$

Until now we have not made use of the convexity of G. As it happens, the convexity of G guarantees the following very important property of the new length:

THEOREM 2.1 (The triangle inequality). *In any triangle ABC, the length of any one side (measured with respect to the figure G) does not exceed the sum of the lengths of the other two sides.*

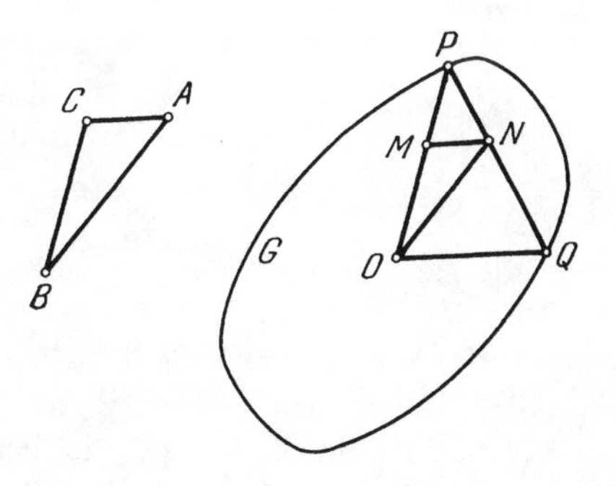

Fig. 2.7

Proof. Let

$$d_G(C, B) = a, \qquad d_G(A, C) = b, \qquad d_G(A, B) = c\,.$$

Draw two "radii" of G, OP and OQ, having the same directions as the vectors **BC** and **CA** (fig. 2.7). Next, take the point M on the segment OP such that $OM/MP = a/b$, and draw (inside the triangle OPQ) the segment MN parallel to OQ. In view of the similarity of the triangles OPQ and MPN, we have

$$d_G(O, M) = OM/OP = \frac{a}{a+b},$$

$$d_G(M, N) = MN/OQ = MP/OP = \frac{b}{a+b}\,.$$

Consequently,

$$BC/OM = d_G(B, C)/d_G(O, M) = a\bigg/\left(\frac{a}{a+b}\right) = a + b\,,$$

$$CA/MN = d_G(C, A)/d_G(M, N) = b\bigg/\left(\frac{b}{a+b}\right) = a + b\,.$$

Thus $BC/OM = CA/MN$; moreover, $\angle BCA = \angle OMN$. This implies that the triangles BCA and OMN are similar, and therefore AB is parallel to NO and $AB/NO = a + b$; that is, $d_G(A, B)/d_G(N, O) = a + b$. Thus,

$$d_G(N, O) = \frac{d_G(A, B)}{a+b} = \frac{c}{a+b}\,.$$

But the points P and Q belong to G, which, because it is *convex*, contains the entire segment PQ. In particular, the point N belongs to G. Hence $d_G(O, N) \leq 1$, which tells us that $c/(a + b) \leq 1$, or, finally, that $c \leq a + b$. And this means that

$$d_G(A, B) \leq d_G(B, C) + d_G(A, C)\,,$$

the statement of the triangle inequality.

A plane in which the units of length are given by some convex centrally symmetric figure G is called a *Minkowski plane*. The figure G is called the *unit disk* of the Minkowski plane.

Let r be a nonnegative real number, and let C be an arbitrary point of a given Minkowski plane. The set of all points A whose distance from

C does not exceed r—that is, the set of points A satisfying the condition $d_G(C, A) \leq r$—is called the *disk of radius r about* C. Note that if two points A and B belong to some disk of radius r, then the distance $d_G(A, B)$ between them cannot exceed $2r$, for if C is the center of this disk, the triangle inequality yields

$$d_G(A, B) \leq d_G(A, C) + d_G(B, C) \leq r + r = 2r .$$

In order to help visualize a disk in the Minkowski plane, we recall the definition of a *dilation*. Let F be a plane figure, and choose an arbitrary point O in the plane and a positive number k. For each point A of the figure F, we find the point A' on the ray OA such that $OA'/OA = k$ (fig. 2.8). The set of all points A' thus obtained forms a new figure F'.

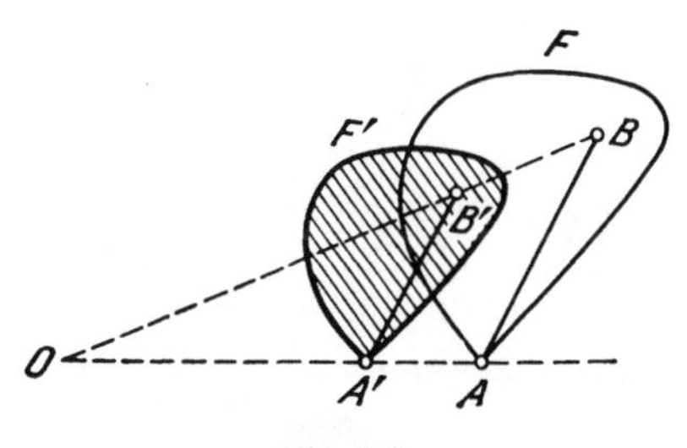

Fig. 2.8

The mapping from the figure F to the figure F' is called a *dilation* with center O and coefficient k, and the figure F' is said to be a *dilation* of F. If F is convex, then a dilation F' of F is also convex (for if a segment AB is entirely contained in F, then its image, the segment $A'B'$, is entirely contained in F').

Note that if the coefficient of dilation k is less than 1, then the figure F' is a shrunken copy of F, and if $k > 1$, then F' is a magnified copy.

We are now in a position to describe all disks in a given Minkowski plane: *a figure is a disk of radius r in the Minkowski plane generated by the unit disk G if and only if it is the image of G under a dilation with coefficient r.* The proof of this assertion is not complicated, and we leave it to the reader.

Let F be a figure in the Minkowski plane with unit disk G. As in ordinary geometry, the *diameter* of F (cf. p. 1) is the *greatest* of the distances between pairs of points of F, that is, the largest of the numbers $d_G(A, B)$ where A and B are any points of F (compare remark 1).

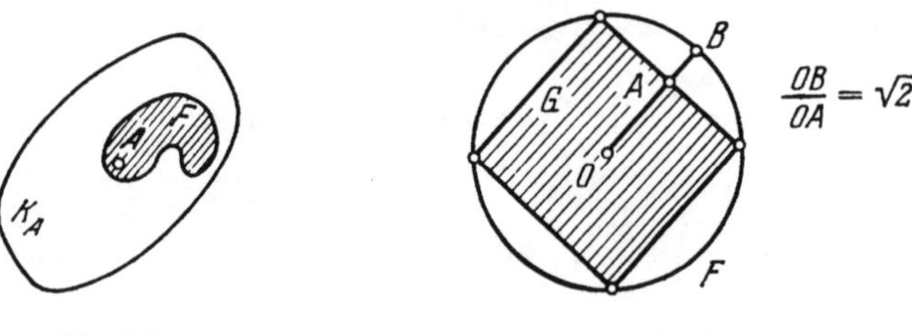

Fig. 2.9 Fig. 2.10

If the diameter of a figure F is less than or equal to d, then of course the disk K_A of radius d centered at an arbitrary point A of F completely contains F (fig. 2.9). Conversely, if every such disk contains F, then the diameter of F does not exceed d.

Consider the example of the Minkowski plane whose unit disk is the square pictured in figure 2.5. It is easy to see that the circumscribed "ordinary" (circular) disk (defined by the inequality $x^2 + y^2 \leq 1$, and consequently having diameter 2 in ordinary geometry) has diameter $2\sqrt{2}$ in this Minkowski geometry (fig. 2.10).

Now let l be a line, and let A be a point not on l. Considering disks of various radii centered at A, we can pick only one of them for which the line l will be supporting. The radius r of this disk is called the *distance from A to l*. The use of this term is justified by the fact that if B is any point of l, B either lies outside the chosen disk (fig. 2.11), in which case

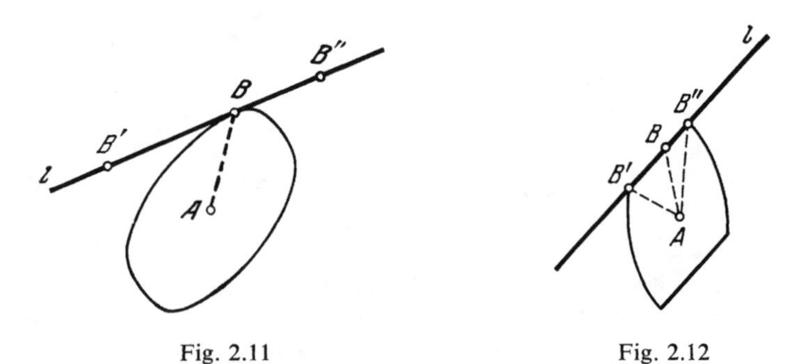

Fig. 2.11 Fig. 2.12

$d_G(A, B) > r$, or lies on the boundary of this disk, in which case $d_G(A, B) = r$. Thus the distance from a point A to a line l is (as in ordinary geometry) the *least* of the distances from A to points of l.

Note that in a Minkowski geometry the distance from a point A to a line l is not, generally speaking, measured along the perpendicular from A to l (fig. 2.11). Moreover, it is possible that a line, instead of having a unique point nearest to A, contains an entire segment of "nearest points" (fig. 2.12). It is easy to see that if l and m are parallel lines, then the distance to m from any point A of l is independent of the position of A on l (and is equal to the distance to l from any point of m). This invariant distance is called the *distance from the line l to the line m*. If, for example, l and m are two parallel supporting lines to the unit disk G, then the distance between them is 2 (fig. 2.13).

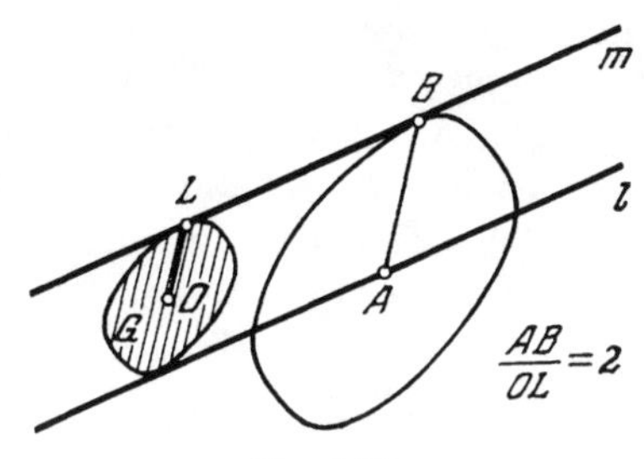

Fig. 2.13

Let F be any convex figure in the Minkowski plane with unit disk G, and let l and m be a pair of parallel supporting lines to this figure. The distance between l and m is called the *width* of F in the direction l. The diameter of any figure is the figure's greatest width (see remark 13).

For example, let the square of figure 2.5 serve as the unit disk G, and let F be the ordinary (circular) disk circumscribed around the square G. Then the width of F in a direction parallel to a side of the square G is $2\sqrt{2}$, and the width of F in a direction parallel to a diagonal of G is 2. Thus, in this Minkowski geometry the circular disk is no longer a figure of constant width. In general, if (in the Minkowski plane with unit disk G) a figure F has the same width in all directions, then F is called a *figure of constant width* in this plane. If, for example, the unit disk G is a regular hexagon, then the equilateral triangle in figure 2.14 is a figure of constant width.

It is curious to note that *any* bounded convex figure containing interior points (that is, which is not a segment or single point) is a figure of constant width in some unique Minkowski geometry (here, of course, we identify Minkowski geometries whose unit disks are dilations of one another) (see remark 14).

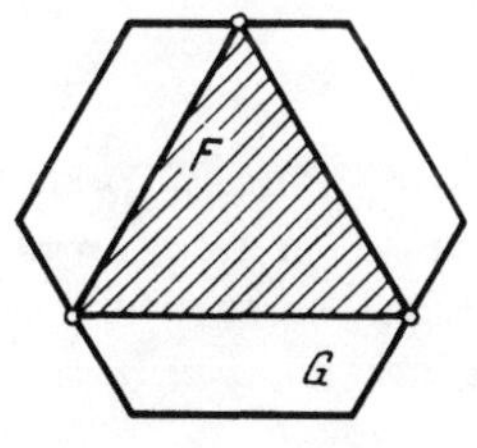

Fig. 2.14

It is clear from the above that all of the definitions considered in sec. 1.4 and 1.5 (diameter of a convex figure, width, figures of constant width) carry over to Minkowski planes. It therefore makes sense to consider the problems analyzed in the first chapter in the context of Minkowski planes, and we now do so.

2.3. Borsuk's Problem in Minkowski Planes

Let F be a figure in the Minkowski plane with unit disk G. Denote the diameter of F in this geometry by d. We shall consider the problem of dividing F into pieces of smaller diameter; the smallest number of pieces required to do so will be denoted by $a_G(F)$. Clearly, the diameters of pieces of the figure, as well as the diameter of the whole figure, depend in an essential way on *which* Minkowski plane is considered (that is, on the choice of the unit disk G). For this reason the number $a_G(F)$ also depends heavily on the choice of the unit disk G.

For example, a parallelogram can be divided into two pieces of smaller diameter in the ordinary plane (fig. 1.12*b*); but if this parallelogram is considered in "its" Minkowski plane (that is, in the Minkowski plane for which it is the unit disk), the diameter of the whole parallelogram and of each of the indicated pieces is clearly 2 (the length of each side and each diagonal of the parallelogram G is 2). It turns out that it is impossible to divide a parallelogram into three pieces of smaller diameter in its own Minkowski plane; four pieces, however, are sufficient for such a division (fig. 2.15).

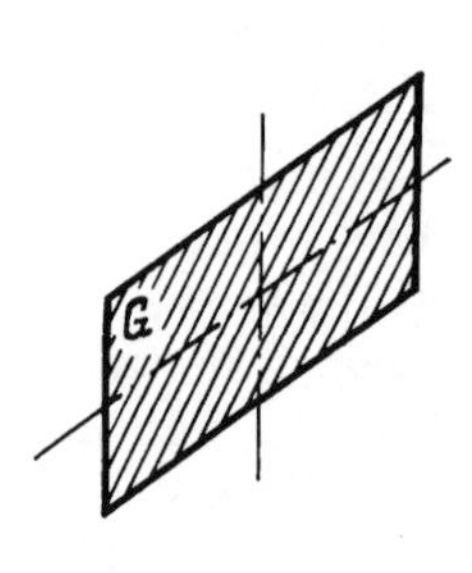

Fig. 2.15

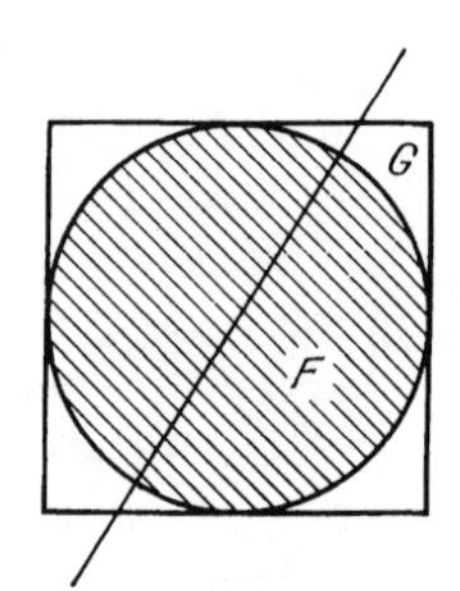

Fig. 2.16

In this case, in other words, $a_G(G) = 4$. This shows that for some figures F and G the inequality $a_G(F) > a(F)$ is possible. There are, however, cases in which $a_G(F) < a(F)$. Indeed, if G is a square and F is a circular disk, it is easy to see that $a_G(F) = 2$ (fig. 2.16), whereas $a(F) = 3$.

The problem of evaluating $a_G(F)$ was studied in 1957 by the American geometer Grunbaum, who proved a theorem similar to the following:

THEOREM 2.2. *For any bounded plane figure F,*

$$a_G(F) \leq 4;$$

equality is attained if and only if G is a parallelogram and the convex hull F is a parallelogram which is a dilation of G (that is, if F contains four points which are the vertices of a parallelogram which is a dilation of G with coefficient of dilation $d/2$, where d is the diameter of F).

We shall prove this theorem later (in sec. 3.4).

The results of sec. 1.7 also have interesting generalizations to Minkowski planes.

Let us first examine the case where the unit disk of the given Minkowski plane is a parallelogram. In this case we have the following theorem:

THEOREM 2.3. *Let F be a figure of diameter d in a Minkowski plane whose unit disk is a parallelogram G. Then*

$$a_G(F) = 2$$

if and only if F does not contain three points which are the vertices of an equilateral triangle with sides of length d.

It is understood, of course, that "equilateral" means equilateral in the given Minkowski metric.

Proof. Suppose F does contain all three vertices of some equilateral triangle with side d. It is clear that no set of diameter less than d can contain any two of these points, hence $a_G(F) \geq 3$.

Now suppose that F does not contain the vertices of any equilateral triangle of side d. Draw four supporting lines l_1, l_2, m_1, and m_2 to the figure F parallel to the sides of the parallelogram G (fig. 2.17). These lines form a parallelogram $ABCD$ containing F. Since the sides of this

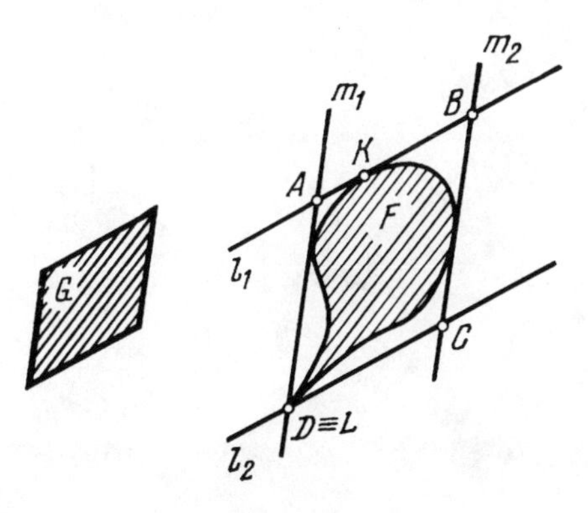

Fig. 2.17

parallelogram are supporting lines of F, each of the sides contains at least one point of F.

Let K and L be points of F lying on the segments AB and CD, respectively. Since both K and L belong to F, $d_G(K, L) \leq d$. Thus, the distance between some point on the line l_1 and some point on the line l_2 is less than or equal to d, and so the distance between l_1 and l_2 does not exceed d. Similarly, the distance between the lines m_1 and m_2 cannot exceed d.

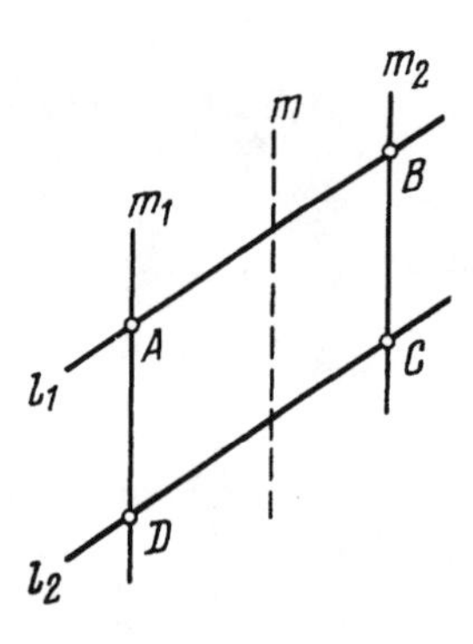

Fig. 2.18

Assume that the distance between l_1 and l_2 is strictly less than d; then, drawing a line m parallel to m_1 and m_2 which crosses the parallelogram $ABCD$ (fig. 2.18), we cut $ABCD$ (and consequently F, which is contained in it) into two pieces, each of diameter less than d. Thus, in this case, $a_G(F) = 2$. Similarly, $a_G(F) = 2$ if the distance between the lines m_1 and m_2 is strictly less than d. It remains for us to examine the case when the distances between l_1 and l_2 and between m_1 and m_2 are both d; that is, when parallelogram $ABCD$ is a disk of radius $d/2$.

Assume first that neither A nor C belongs to F. Then, as we can easily see, the line AC divides F into two parts, each of diameter less than d. Indeed, in this case one may draw lines n_1 and n_2 parallel to the diagonal BD and thereby carve from the parallelogram $ABCD$ a hexagon that contains F (fig. 2.19). The line AC cuts this hexagon, and thus its subset

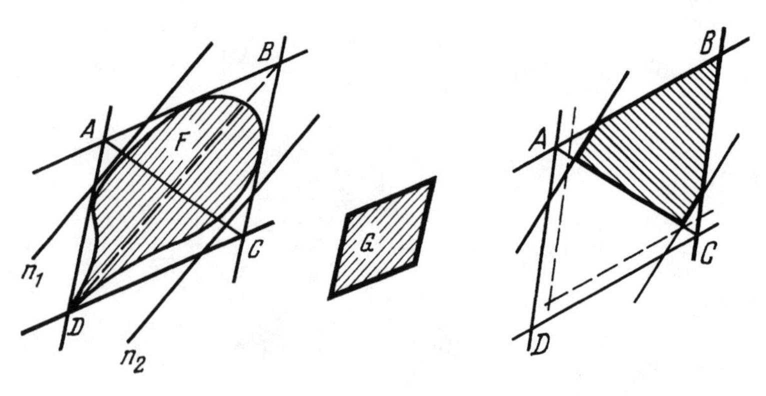

Fig. 2.19 Fig. 2.20

F, into two parts, each of which has diameter less than d (fig. 2.20). Thus $a_G(F) = 2$.

The same proof shows that if neither B nor D belongs to F, then $a_G(F) = 2$.

The remaining case is that in which at least one of the points A or C and at least one of the points B or D belong to F. But this case is impossible; suppose that at least one of the points A or C (say A) belongs to F, and at least one of the points B or D (say B) belongs to F. Let L be a point of the segment CD belonging to F (recall that CD is a supporting line of this figure). Then (see fig. 2.21)

$$d_G(A, B) = d_G(A, L) = d_G(B, L) = d;$$

that is, ABL is an equilateral triangle with side d. But this contradicts the assumption that F does not contain any three points which are the vertices of an equilateral triangle of side d.

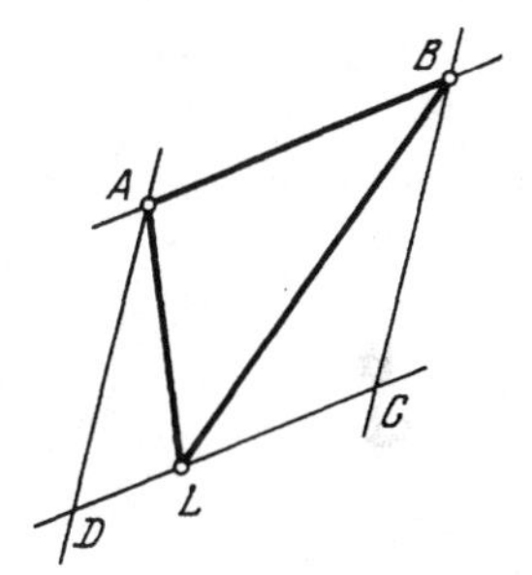

Fig. 2.21

The theorem is thus proved.

Combining theorems 2.2 and 2.3, we make the following assertion, giving a complete solution to the problem of determining the value of $a_G(F)$ when G is a parallelogram:

If a figure F of diameter d contains four points which are the vertices of some parallelogram which is a dilation of G with coefficient $d/2$ (fig. 2.22), *then* $a_G(F) = 4$; *if this is not the case, but F contains the three vertices of some equilateral triangle of side d* (fig. 2.23), *then* $a_G(F) = 3$; *otherwise* $a_G(F) = 2$.

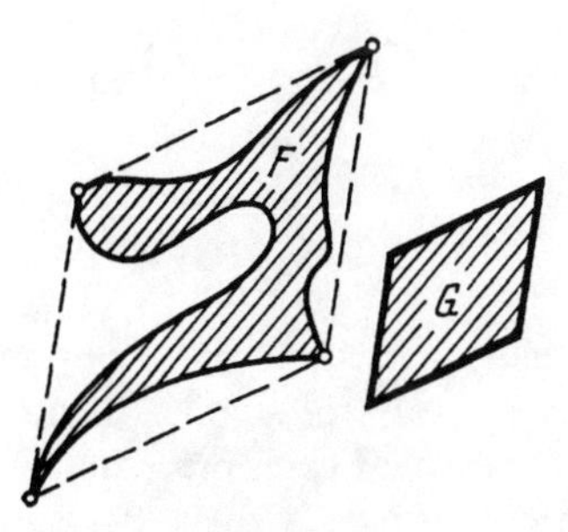

Fig. 2.22

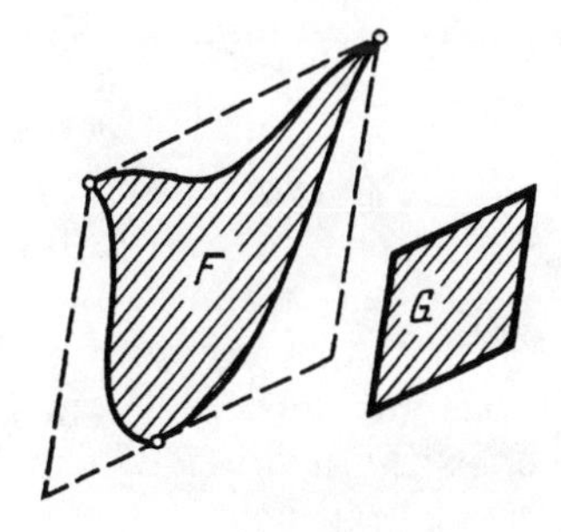

Fig. 2.23

Thus, the problem of completely determining the quantity $a_G(F)$ is solved in two cases: for ordinary geometry (that is, when G is circular) in sec. 1.7, and for the case where G is a parallelogram in this section. A complete solution is not known to the authors; it is known only (see theorem 2.2) that $2 \leq a_G(F) \leq 3$ if G is not a parallelogram. The following conjecture seems likely to us:

Let F be a figure of diameter d in a Minkowski plane with a unit disk G which is not a parallelogram. Then the following two conditions are necessary and sufficient for $a_G(F) = 3$:

1. *F has a unique completion to a figure of constant width in the given Minkowski plane.*

2. *If F^* is the extension of F (that is, the intersection of all disks of radius d containing F), then in each pair of parallel supporting lines of F^* separated by a distance of d, at least one touches F.*

Note that if the figure G is an ordinary circular disk, then the second condition is redundant; one can show that it follows from the first. In the general case, however, condition (2) is necessary. One can become convinced of this through the following example: let F be an ordinary (circular) disk, and let G be the figure depicted in figure 2.24. Then, as can be seen, F has a unique completion to a figure of constant width; the only figure of constant width (and the same diameter) containing F is G, so condition (1) is fulfilled. But condition (2) is not, and $a_G(F) = 2$ (see the dotted line in fig. 2.24). Thus condition (1) alone is not sufficient in general to assure that $a_G(F) = 3$. This complication is directly related to the fact that $a_G(F)$ does not in general coincide with $a_G(F^*)$.

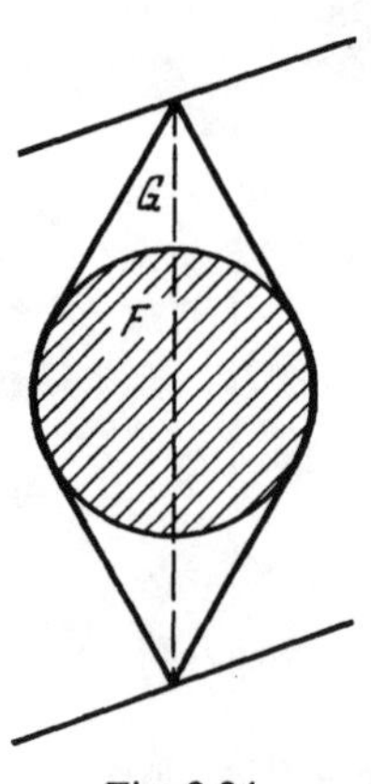

Fig. 2.24

We have deferred the proof of theorem 2.2, which gives an upper bound for $a_G(F)$. One of this theorem's corollaries is a description of the case $F = G$: $a_G(G) = 4$ if G is a parallelogram, and $a_G(G) \leq 3$ in all other cases. But $a_G(G) = 2$ is impossible. (This is proved in the same way as in the case when G is a circular disk; see page 3.) Thus, if G is not a parallelogram, $a_G(G) = 3$.

For the cases in which G is a circular disk or a parallelogram, divisions into three and four pieces (respectively) of smaller diameter are shown in figures 2.25 and 2.15. In these cases each piece not only has a smaller diameter but also can be covered by a disk of smaller diameter in the given Minkowski plane. One might naturally speculate that this is true of any Minkowski plane. In other words, the following conjecture arises:

Any convex centrally symmetric figure G other than a parallelogram can be divided into three parts, each of which can be covered by a dilation of G with coefficient of dilation less than 1 (that is, by a "disk" of smaller diameter).

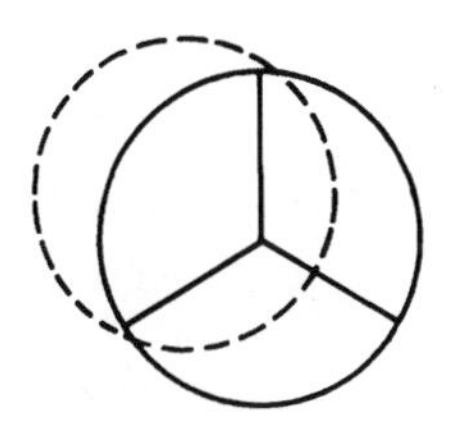

Fig. 2.25

This conjecture has actually been proved, and not only for centrally symmetric figures, but for *all* convex figures. The next chapter is devoted to the proof of this fact and, based on it, the proof of theorem 2.2.

3 The Covering of Convex Figures by Reduced Copies

3.1. Formulation of the Problem

In this chapter we shall consider the problem, given a bounded convex plane figure F, of finding the minimal number of *reduced copies* of F needed to cover the whole figure F. We shall denote this minimal number by $b(F)$. More precisely, the equation $b(F) = m$ means that there exist dilations $F_1, F_2, \ldots, F_m$ of F with coefficients of dilation less than 1, the union of which covers all of the figure F (fig. 3.1), and that this number m is minimal; that is, that no smaller number of dilations of F with coefficients less than 1 is sufficient for this purpose.

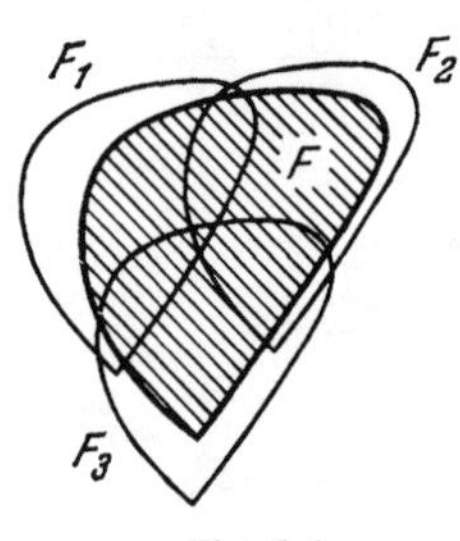

Fig. 3.1

The problem of what values $b(F)$ can assume was posed in 1960 by I. Ts. Gokhberg and A. S. Markus. This problem was considered somewhat earlier (but in a different form) by the German mathematician Friedrich Levi.

As an example, let us consider the case where F is a circular disk. Then any disk of diameter smaller than that of F is a reduced copy of F. It is fairly evident that it is impossible to cover the original disk F with two such disks; that is, that $b(F) \geq 3$. Indeed, suppose F_1 and F_2 (with centers O_1 and O_2) are two disks of smaller diameter (fig. 3.2). Drop a perpendicular from the center O of the original disk F to the line O_1O_2. This perpendicular intersects the circumference of F at two points A and B. Suppose, for example, that A is on the same side of the line O_1O_2 as O. (If O_1O_2 passes through O, we may take either of the two points A and B.) Then $AO_1 \geq AO = r$ and $AO_2 \geq AO = r$, where r

is the radius of F. Since the disks F_1 and F_2 have radii less than r, the point A cannot belong to either of them; that is, the disks F_1 and F_2 do not cover the whole disk F.

On the other hand, it is easy to cover F with three disks of somewhat smaller diameter (fig. 3.3). Thus for a disk F, $b(F) = 3$.

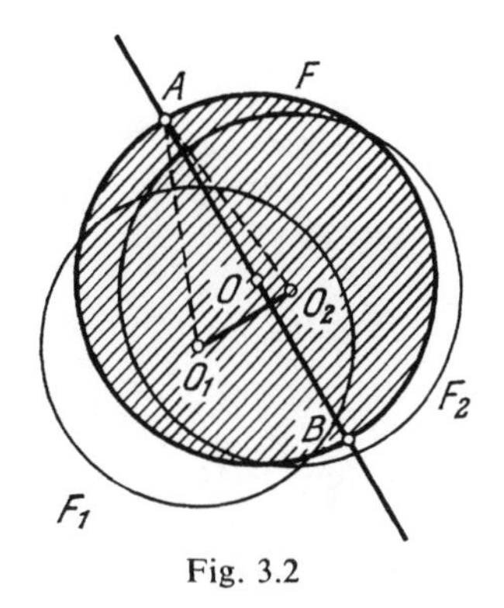

Fig. 3.2

Let us consider another case, that of a parallelogram. Clearly no single reduced copy of F with coefficient less than 1 can contain two vertices of F. In other words, the four vertices of F must belong to four *distinct* reduced copies; that is, $b(F) \geq 4$. And four reduced copies are clearly sufficient (fig. 3.4). Thus, for a parallelogram, $b(F) = 4$.

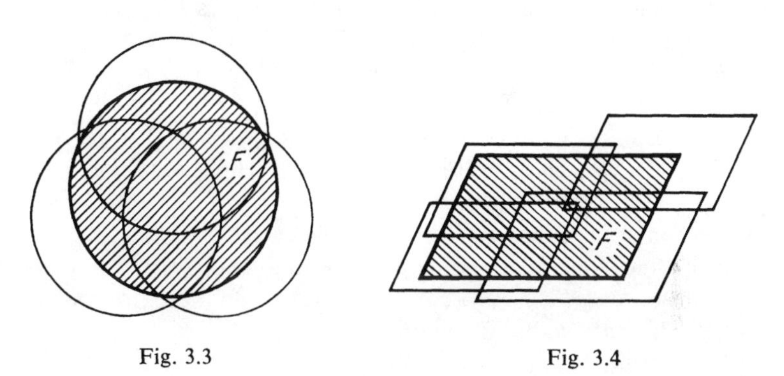

Fig. 3.3 Fig. 3.4

3.2. Another Formulation of the Problem

The problem of covering a figure with reduced copies of itself can be formulated differently, bringing it closer to Borsuk's problem of dividing a figure into pieces of smaller diameter.

Suppose G is a figure contained in the convex figure F. We shall say that the piece G of the figure F has *bulk k* if there exists a dilation F' of F with coefficient k containing G, but no dilation of F with coefficient less than k completely containing G (see remark 15). Obviously, if the piece G coincides with the figure F, its bulk is 1. Of course, any piece which does not coincide with F has bulk less than or equal to 1, but not necessarily strictly *less than* 1. For example, if F is a disk and G is an acute triangle inscribed in F (fig. 3.5), then the bulk of the piece G is 1

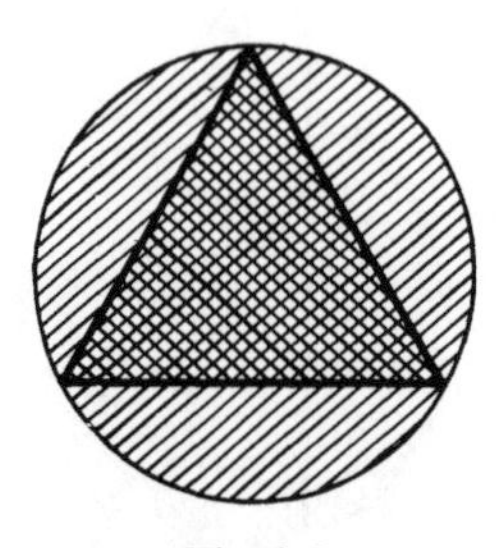

Fig. 3.5

(because no disk of smaller diameter completely contains G). We shall call a figure G contained in a figure F a *piece of a smaller bulk* if its bulk is strictly less than 1.

Using the concept of bulk, we can reformulate the definition of the quantity $b(F)$: *$b(F)$ is the minimal number of pieces of smaller bulk into which the given convex figure F can be divided.* It is readily apparent that this definition is equivalent to the previous one, for suppose $F_1, F_2, \ldots, F_m$ are reduced copies of F whose union covers F. Denote by $G_1, G_2, \ldots, G_m$ the pieces of F cut out by the figures $F_1, F_2, \ldots, F_m$. It is evident that each of these pieces has bulk less than 1. Thus, if a figure F can be covered with m reduced copies, then it can be divided into m pieces of smaller bulk. Conversely, if F can be divided into m pieces $G_1, G_2, \ldots, G_m$ of smaller bulk, then there exist reduced copies $F_1, F_2, \ldots, F_m$ of F containing the pieces $G_1, G_2, \ldots, G_m$, respectively. These figures $F_1, F_2, \ldots, F_m$ form a covering of F by reduced copies of itself.

Thus, the problem of covering a convex figure by reduced copies of itself can also be formulated as the *problem of dividing the figure F into pieces of smaller bulk*. In this form it is strongly reminiscent of the problem of Borsuk studied in chapter 1.

The connection between these problems is more than superficial. Indeed, if the figure F has diameter d, then a dilation of F with coefficient k has diameter kd. Therefore, if a piece of a convex figure has smaller bulk, it also has smaller diameter. (The converse, generally speaking, is false; for example, an equilateral triangle inscribed in a disk F is a piece of smaller diameter, but its bulk is 1; see fig. 3.5.) Thus, if a convex figure F can be divided into m pieces of smaller bulk, then it can certainly be divided into m pieces of smaller diameter (but not conversely, as the example of the parallelogram shows).

Thus, for any convex figure F, we have

$$a(F) \leq b(F)\,.$$

Note that the problem of division into pieces of smaller bulk concerns only *convex* figures, whereas Borsuk's problem of division into pieces of smaller diameter is stated for all figures.

3.3. Solution of the Covering Problem

As we have seen in sec. 3.1, when we consider the problem of covering a convex figure with reduced copies, the circular disk is no longer the figure requiring the largest number of covering figures; $b(F)$ has a greater value for a parallelogram than for a disk.

One question immediately arises: do there exist convex plane figures for which $b(F)$ has a value still greater than that for a parallelogram? It turns out that the answer is no; in fact, among all convex plane figures, $b(F) = 4$ *only* for parallelograms. In other words, we have the following theorem, established in 1960 by I. Ts. Gokhberg and A. S. Markus (somewhat earlier, in 1955, Friedrich Levi obtained a different result, from which this theorem can be deduced):

THEOREM 3.1. *For any bounded convex plane figure F other than a parallelogram,* $b(F) = 3$; *if F is a parallelogram, then* $b(F) = 4$.

Before proving theorem 3.1, we shall introduce one definition and prove several auxiliary propositions.[1]

Let F be a convex figure, and let A and B be two of its boundary points. We shall call the segment AB a *major chord* of F if every chord parallel to and distinct from AB is strictly shorter than AB. For example, in a disk all the diameters, and only the diameters, are major chords (fig. 3.6). In the case of a parallelogram, the only major chords are the diagonals (fig. 3.7). In general, if l_1 and l_2 are two parallel supporting

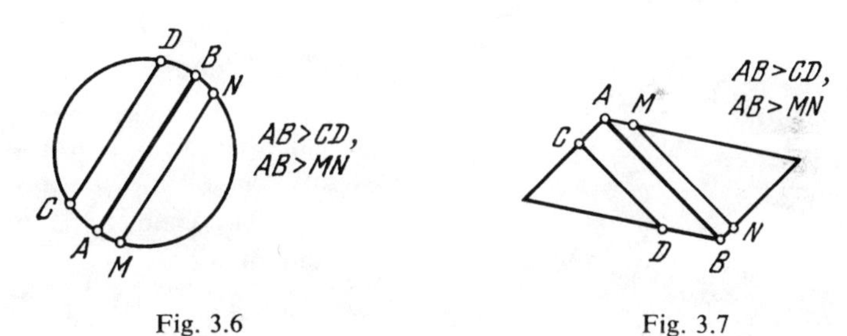

Fig. 3.6 Fig. 3.7

lines to the convex figure F, each of which has only one point of contact with F, then the segment connecting the points of contact is a major chord of F (fig. 3.8). If one of the parallel supporting lines l_1, l_2 intersects F in a unique point A and the other has a segment BC in common with F, then the segment connecting A with any point D of the segment BC is a major chord of F (fig. 3.9). Finally, if both of the intersections of the two supporting lines l_1 and l_2 with F are segments, then both diagonals of the trapezoid defined by the endpoints of these segments

1. Altogether (including the auxiliary propositions) the proof of theorem 3.1 occupies 10 pages. Later (in sec. 4.3) another proof will be presented.

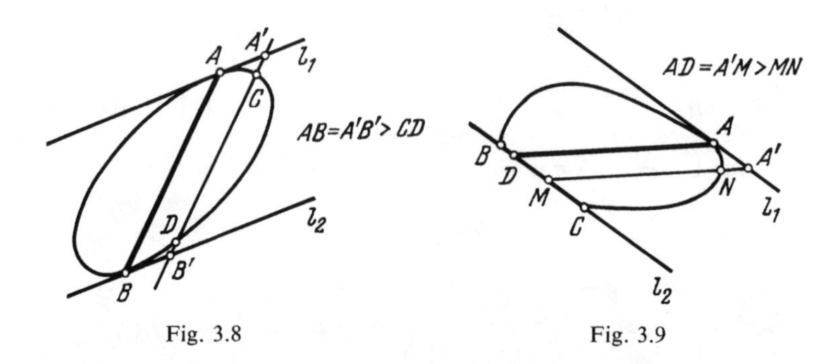

Fig. 3.8

Fig. 3.9

are major chords of F (fig. 3.10). It follows from what has been said that *if l_1 and l_2 are two parallel supporting lines to the convex figure F, then at least one segment with end-points on these lines is a major chord of F.*

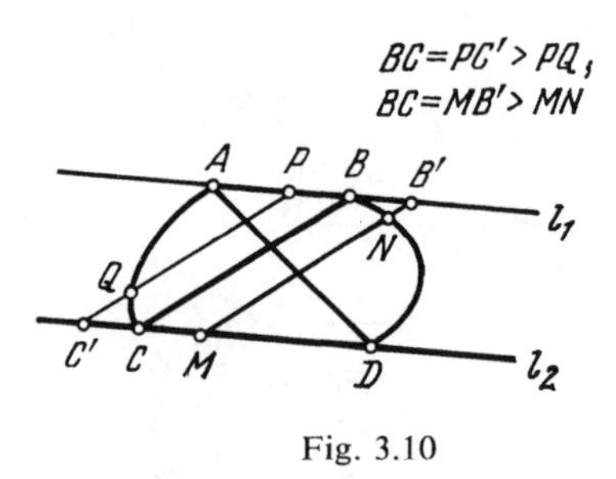

Fig. 3.10

THEOREM 3.2. *Any convex figure other than a parallelogram has at least three major chords.*

In other words, parallelograms are the *only* convex figures with only two major chords. A regular hexagon is an example of a figure with exactly three major chords. A regular $2n$-gon has exactly n major chords, and a circular disk has infinitely many.

Proof of Theorem 3.2. Suppose first that the boundary of the convex figure F contains no line segments. Draw two parallel supporting lines l_1 and l_2 to F, and let A and B be the points at which these lines meet F. Then AB is a major chord. Let C be any boundary point of F distinct from A and B. Through the point C draw a supporting line m_1 to F_1, and let m_2 be the supporting line parallel to m_1; denote by D the point at which m_2 meets F. (The possibility that D might coincide with A or B is not excluded.)

The segment CD is a second major chord of F (fig. 3.11). Finally, take any boundary point E distinct from A, B, C, and D. Now, drawing the supporting line n_1 through E and the parallel supporting line n_2, we obtain still another major chord. Thus, in the case under consideration we have found three major chords. (In fact, the same line of reasoning clearly shows that in this case F has an infinite number of major chords.)

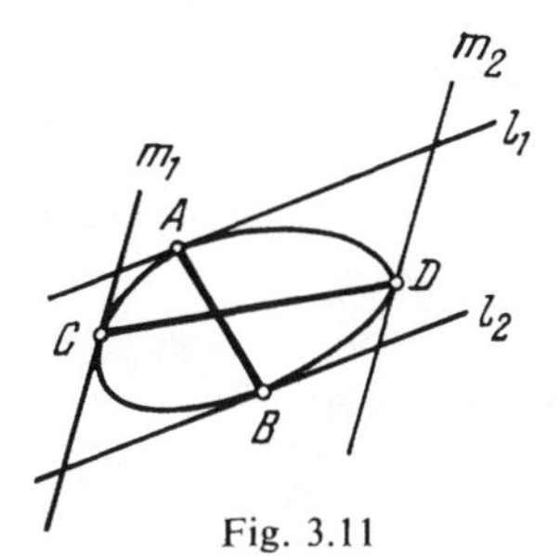

Fig. 3.11

Now assume that the boundary of F contains at least one line segment—that is, that some supporting line l_1 intersects F in a segment AB. Let l_2 denote the supporting line parallel to l_1. If l_2 touches F only at one point C, then any segment connecting C with any point of the segment AB is a major chord. So in this case, too, we have an infinite number of major chords (fig. 3.12).

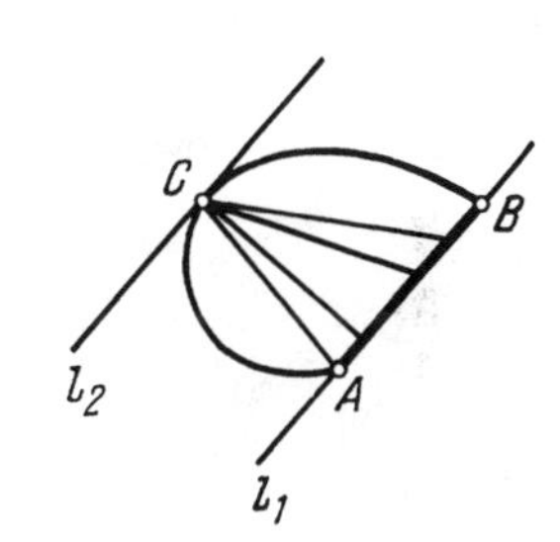

Fig. 3.12

Finally, suppose that the line l_2 also has a segment DC in common with F (fig. 3.13). Then the convex figure F must contain the whole trapezoid $ABCD$, and both diagonals of this trapezoid are major chords of F. If F does not coincide with the trapezoid $ABCD$, then it has some supporting line p_1 which does not touch the trapezoid. Drawing the supporting line p_2 parallel to p_1, we obtain a third major chord (with endpoints on the lines p_1 and p_2).

It remains only to consider the case when F coincides with the trapezoid $ABCD$. If $AB \neq CD$, then the longer of the two segments

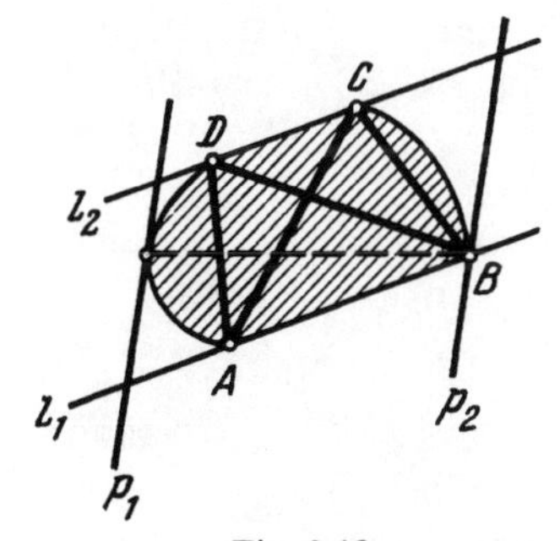

Fig. 3.13

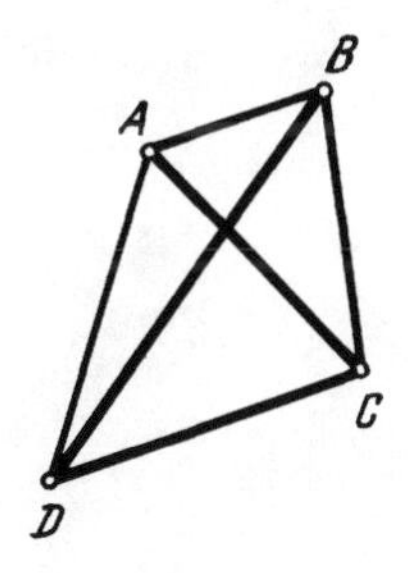

Fig. 3.14

AB, CD is a third major chord of the trapezoid F (fig. 3.14). If $AB = CD$, then F is a parallelogram, and it is evident that a parallelogram has only two major chords.

The theorem is thus proved.

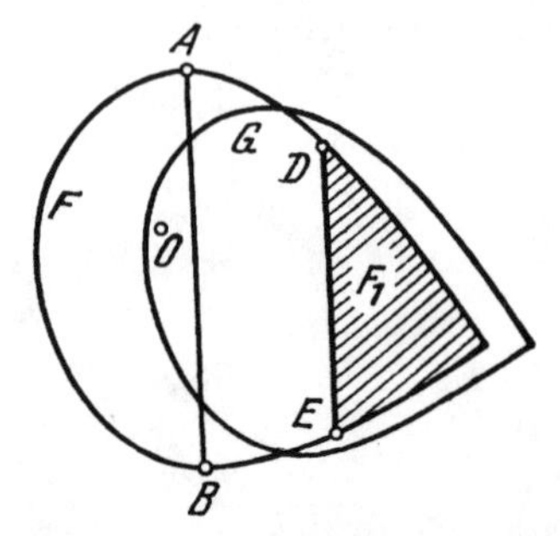

Fig. 3.15

LEMMA 3.1. *Suppose AB is a major chord of a convex figure F, DE is a chord parallel to AB, and O is an arbitrary interior point of F. Denote by F_1 the part of F marked off by the chord DE and not containing the chord AB. Then there exists a reduced copy G of F completely containing the figure F_1 and the point O* (fig. 3.15).

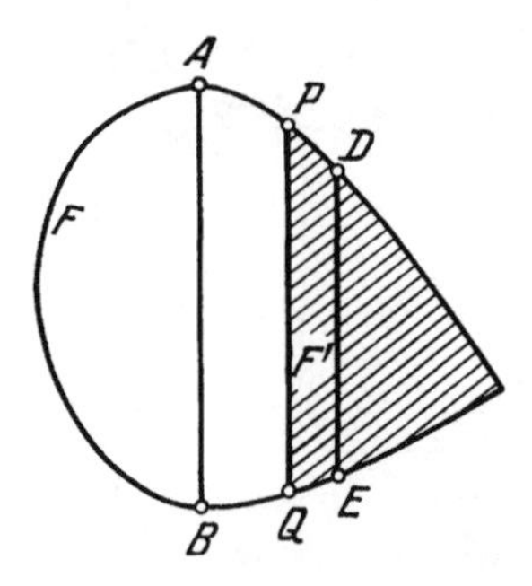

Fig. 3.16

Proof. Let PQ be a chord of F parallel to the chords AB and DE and located between them. Denote by F' the part of F marked off by PQ and containing the chord DE. F' clearly contains F_1 (fig. 3.16). Since AB is a major chord, $AB > PQ$. Let G' be the dilation of F with center at the point T of intersection of the lines AP and BQ (since $AB > PQ$, the lines are not parallel) and with coefficient PQ/AB (fig. 3.17). Under this dilation the chord AB of F is clearly mapped to the chord PQ of G'.

Let M be an arbitrary point of F' not on the chord PQ. Draw lines from the points A and B parallel to PM and QM, respectively, and denote their point of intersection by N (fig. 3.17). The point N belongs

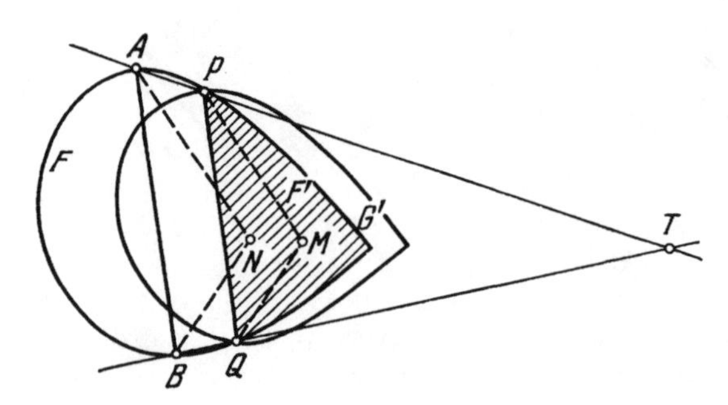

Fig. 3.17

to F since it belongs to the polygon $ABQMP$, which is contained in F (since F is convex). Under the given dilation, the point N of F goes to the point M; it follows that M belongs to G'. Thus, any point M of the figure F' belongs to G'; that is, F' is contained in G'.

No matter which chord PQ we take, provided it is parallel to the chords AB and DE and between them, this construction gives us a reduced copy G' of F containing the figure F' and thus the figure F_1. What remains to be shown is that an appropriate choice of the chord PQ will assure that the point O will also belong to G'; this will yield the desired figure G.

If O is on the same side of AB as DE, then it suffices to take for PQ a chord separating O from the segment AB. Indeed, for such a choice of the chord PQ (fig. 3.18), O belongs to F' and thus is in G'.

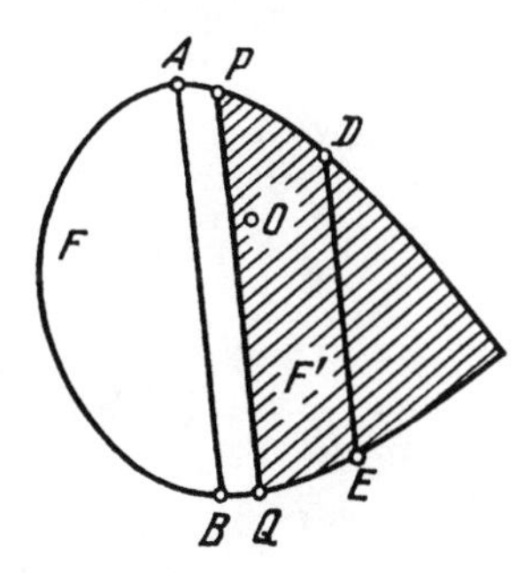

Fig. 3.18

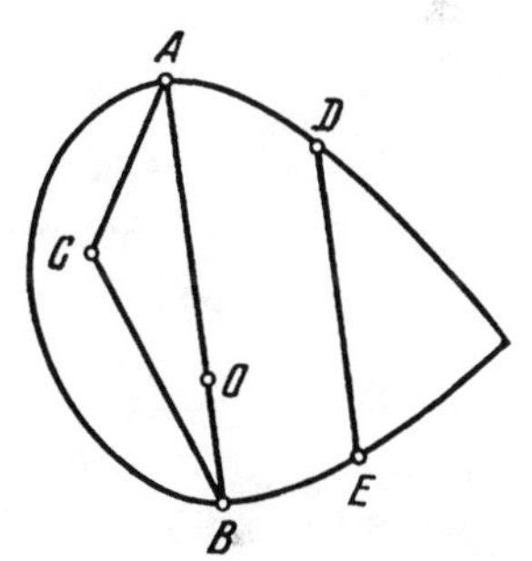

Fig. 3.19

Suppose now that the point O is either on the chord AB or on the opposite side of this chord from segment DE. If O is on AB, then we shall denote by C an arbitrary point of F on the opposite side of AB from DE (fig. 3.19). If O is not on the segment AB, then we draw a ray through O from an arbitrary interior point H of the segment AB, and we let C be the point of intersection of this ray with the boundary of F (fig. 3.20). In both cases O belongs to the triangle ABC but does not lie on the broken line ACB, and the triangle ABC lies on the opposite side of the chord AB from the segment DE.

Draw rays l_1 and l_2 from O in the direction of the vectors $\mathbf{CA}$ and $\mathbf{CB}$, and denote by K and L the points where these rays cross the boundary of F (fig. 3.21). Clearly, the points K and L lie on the same side of AB as does the segment DE. We now choose a chord PQ parallel to AB in such a way that AB will lie on one side of the line PQ and all four points D, K, L, and E will lie on the other side.

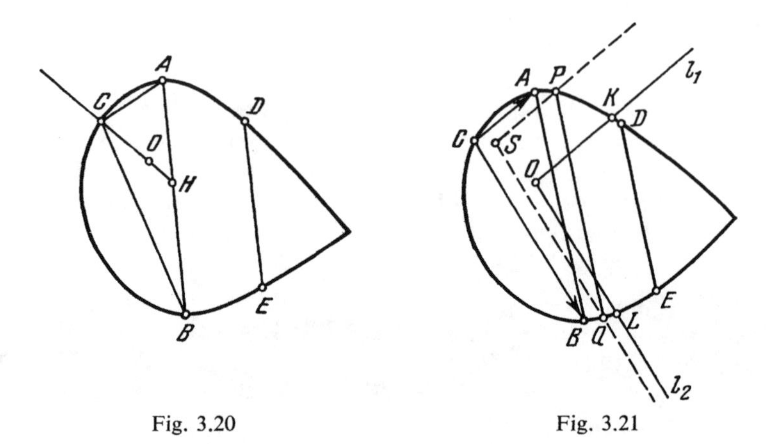

Fig. 3.20 Fig. 3.21

Now draw lines through the points P and Q parallel to the lines AC and CB, respectively, and denote their point of intersection by S. Under the dilation with center T and coefficient PQ/AB, the triangle ABC is clearly mapped to the triangle PQS (recall that T is the point of intersection of the lines AP and BQ). Now we need only show that the point O belongs to the triangle PQS, for this will mean that the point mapped to O by the dilation belongs to the triangle ABC and thus to F, whence it will follow that O belongs to the figure G'.

By the choice of the chord PQ, the points K and L are in the interior of the angle PSQ, and O must consequently lie in the interior of this angle, too (fig. 3.21). Yet O and S lie on the same side of the line PQ. From this it follows that O is an interior point of the triangle PQS.

LEMMA 3.2. *Any two major chords of a convex figure F intersect (in the interior of F or on its boundary).*

Proof. Let AB be a major chord of F, and let CD be a chord of F not intersecting AB. We shall prove that CD is not a major chord. If the chord CD is parallel to AB, this is immediate. Suppose, therefore, that CD is not parallel to AB and, for the sake of definiteness, that the point C is closer to the line AB than is D (fig. 3.22). Draw the chord CM parallel to AB. Since the points B and D are on opposite sides of the line CM, the segment BD must intersect the line CM at some point K. Because F is convex, the point K (which lies on the segment BD) belongs to F. Therefore $CM \geq CK$. At the same time, since AB is a major chord, $CM < AB$; hence $CK < AB$. This implies that the rays AC and BD intersect at some point O (fig. 3.22). Finally, introduce in

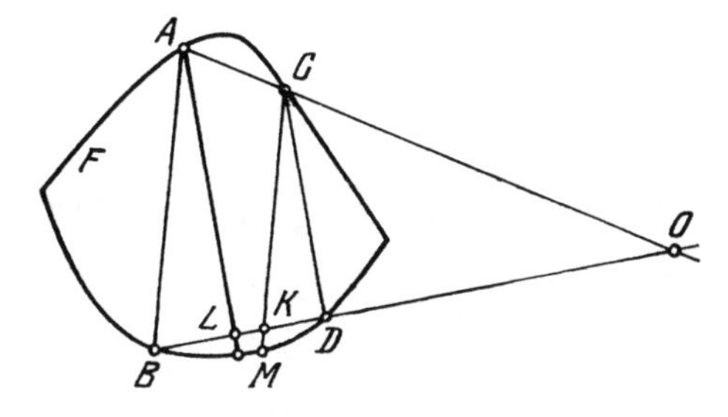

Fig. 3.22

the quadrilateral $ACDB$ the segment AL parallel to CD. The point L belongs to the segment BD and thus to F. But the similarity of the triangles CDO and ALO implies that $AL > CD$. Thus, the line AL crosses F with a chord longer than CD, and therefore CD cannot be a major chord.

We note that since it is easily verified that each diameter of a convex figure is also a major chord, we have proved the assertion of sec. 1.5 (p. 14), namely, that any two diameters of a convex figure intersect at a point of that figure.

Proof of theorem 3.1. First we shall show that if the convex figure F is not a parallelogram, then $b(F) \leq 3$.

By theorem 3.2, the figure F has three distinct major chords, and each pair of them intersects (lemma 3.2). We first consider the case where two distinct major chords AB and AC intersect at the boundary point A. Let O be an arbitrary interior point of F lying in the interior of the angle BAC (fig. 3.23). Draw chords DE and KL through O which are parallel respectively to the chords AB and AC. By lemma 3.1, the part of F bounded by DE and not containing A can be covered by one reduced copy of F. The same conclusion holds for the part of F bounded by the chord KL and not containing A. Furthermore, by choosing the point O sufficiently close to A, one can enclose the remaining curvilinear "sector" OLD in an arbitrarily small disk centered at A. Consequently, this part of F can be covered by one reduced copy of F with an arbitrarily small coefficient.

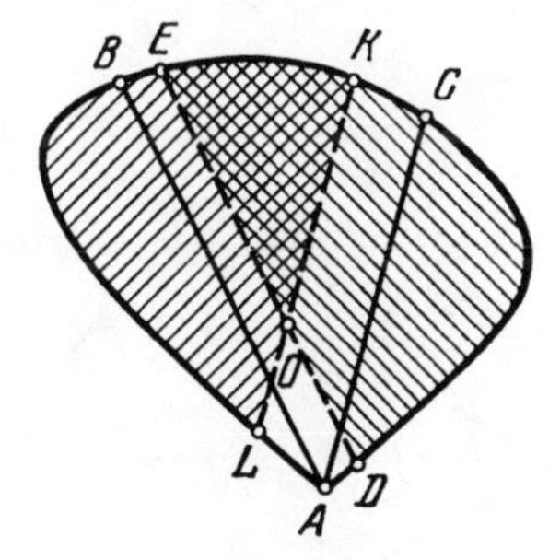

Fig. 3.23

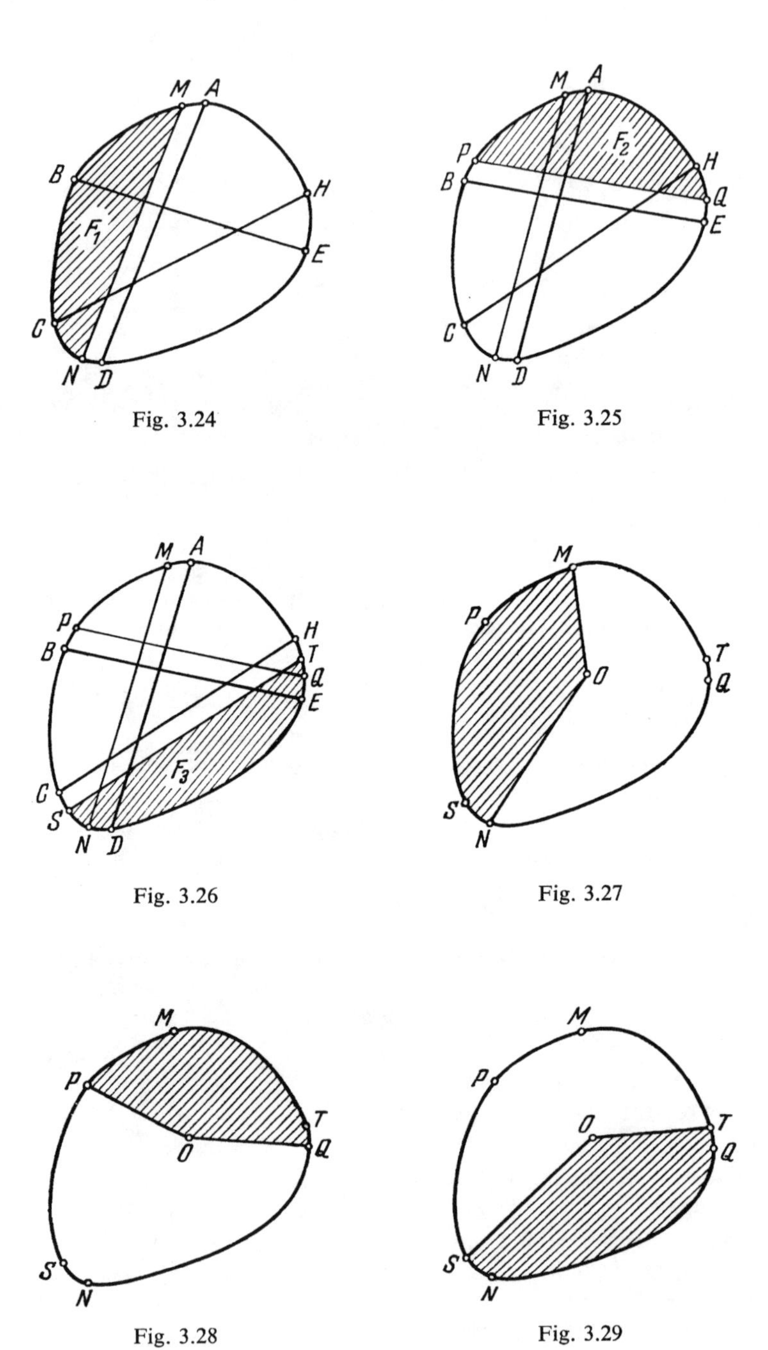

Fig. 3.24

Fig. 3.25

Fig. 3.26

Fig. 3.27

Fig. 3.28

Fig. 3.29

Thus it remains to consider the case where each pair of the three major chords intersects at an interior point of F. Let us denote the six endpoints of these three chords by the letters A, B, C, D, E, and H, going around the curve counterclockwise, so that the chords under discussion will be AD, BE, and CH (fig. 3.24). Let MN be a chord parallel to AD which separates AD from the points B and C (fig. 3.24). Denote by F_1 the part of F marked off by the chord MN which does not contain AD. Next draw a chord PQ parallel to BE which separates BE from the points M, A, and H (fig. 3.25). Denote by F_2 the part of F marked off by this chord which does not contain BE. Finally, draw a chord ST parallel to CH which separates CH from the points N and Q, and denote by F_3 the part of F marked off by this chord which does not contain CH (fig. 3.26).

Now let O be an arbitrary interior point of F. By lemma 3.1, there exist reduced copies G_1, G_2, and G_3 of F containing the figures F_1, F_2, and F_3, respectively, along with the point O. G_1 (as a convex figure) therefore contains the "sector" of F bounded by the broken line MON (fig. 3.27), and the figures G_2 and G_3 contain analogous sectors POQ and SOT (figs. 3.28 and 3.29). The union of the figures G_1, G_2, and G_3, since it contains these three sectors, must contain the figure F.

Thus, in this case too,

$$b(F) \leq 3 .$$

Let us show, finally, that $b(F) \geq 3$ for any convex figure F. Let F_1 and F_2 be two arbitrary reduced copies of F, and let O_1 and O_2 be the corresponding centers of dilation. Draw a line l through points O_1 and

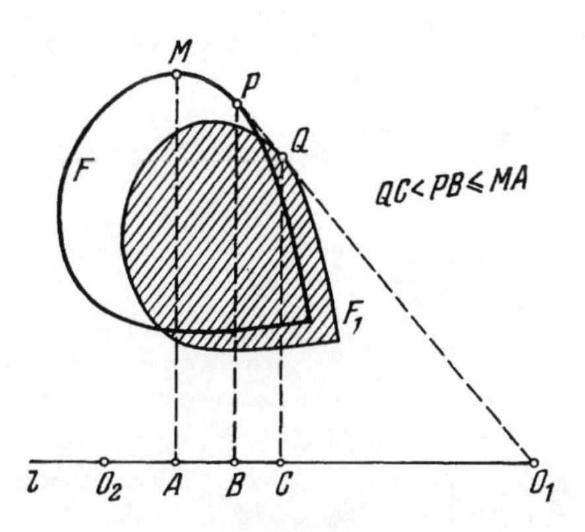

Fig. 3.30

O_2, and let M be a point of F maximally distant from l (fig. 3.30). Then all points of F_1 are closer to l than M is, and the same is true of F_2. Therefore the point M is not contained in either of the figures F_1 or F_2, whence it follows that $b(F) > 2$.

Thus, if F is not a parallelogram, $b(F) = 3$, and for a parallelogram, as we have verified, $b(F) = 4$.

The proof of the theorem is thus complete.

3.4. Proof of Theorem 2.2

We now undertake the task of proving theorem 2.2.

Examining the case when F is convex, let us first observe that the inequality $a(F) \leq b(F)$ retains its validity in any Minkowski plane:

$$a_G(F) \leq b(F) .$$

This is established in exactly the same way as in ordinary geometry (p. 42). Therefore, if F is not a parallelogram, theorem 3.1 implies that $a_G(F) \leq b(F) = 3$.

Now suppose F is a parallelogram. Through the origin O of the Minkowski plane under consideration, we draw two lines parallel to the diagonals of F, and denote by A_1C_1 and B_2D_2 the segments of these lines that lie in G. Using the segments A_1C_1 and B_2D_2 as diagonals, construct two parallelograms with sides parallel to the sides of the parallelogram F (fig. 3.31). Denote by F' the smaller of these and label

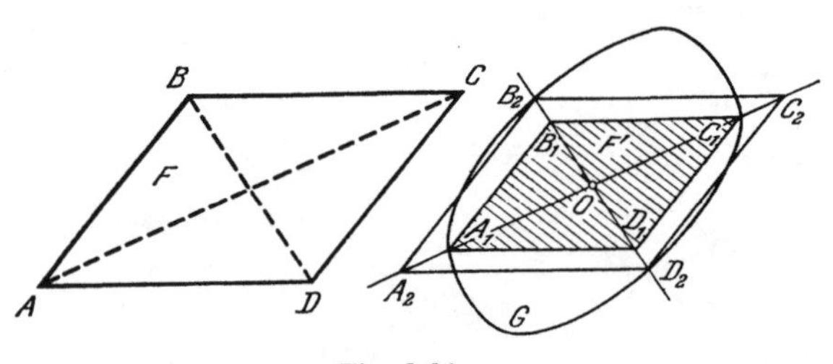

Fig. 3.31

it $A_1B_1C_1D_1$, with diagonal A_1C_1. Since F' is a dilation of F, $a_G(F) = a_G(F')$, so that it suffices to consider the parallelogram F'.

Clearly, $d_G(A_1, C_1)$ is 2, and therefore the diameter of F' (in the Minkowski metric d_G) is at least 2. But the diameter of F' can be no greater than 2, for F' is entirely contained in G, which has diameter 2.

Thus, the diameter of the parallelogram F' is 2.

Now draw a line p through the point O parallel to the sides A_1B_1 and C_1D_1 and denote by M and N the points of its intersection with the boundary of G (fig. 3.32). If the points M and N do not lie on the sides B_1C_1 and A_1D_1 of the parallelogram F' (fig. 3.32), then the entire

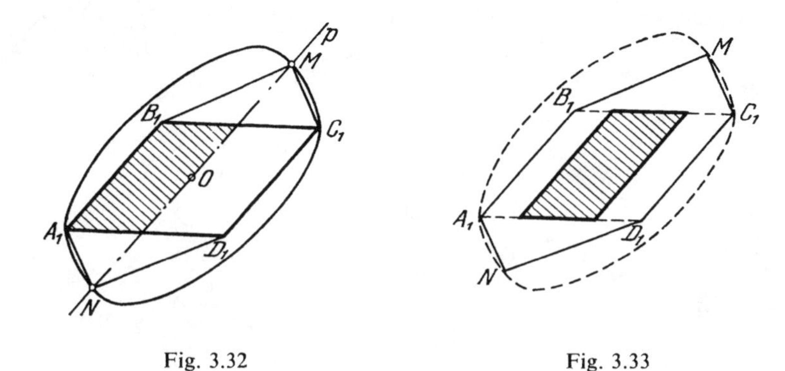

Fig. 3.32 Fig. 3.33

hexagon $A_1B_1MC_1D_1N$ is contained in G. It follows easily that the diameter of each of the "halves" into which the line p divides F' is less than 2 (fig. 3.33), so that $a_G(F') = 2$. If, however, M and N lie on the sides of F', then the lines B_1C_1 and A_1D_1 must be *supporting* lines of G (for there must exist a supporting line through the boundary point M of G, and any line other than B_1C_1 cuts the parallelogram F' and hence

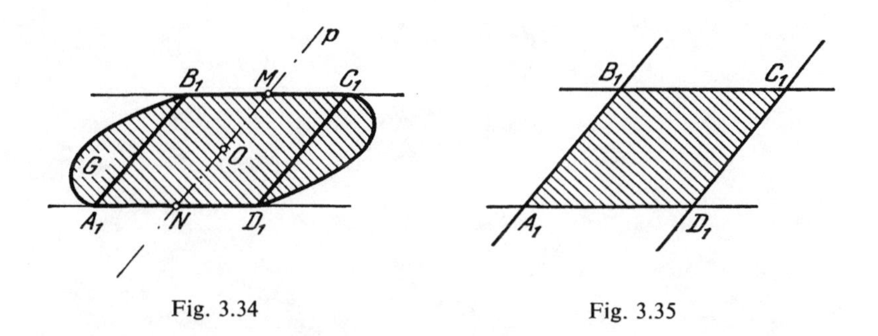

Fig. 3.34 Fig. 3.35

G). Thus (fig. 3.34), the entire figure G lies within the strip between the lines B_1C_1 and A_1D_1.

Consequently, either $a_G(F') = 2$, or the figure G is enclosed in the strip between the lines B_1C_1 and A_1D_1.

Similarly, introducing a line q parallel to the sides B_1C_1 and A_1D_1, we find that either $a_G(F') = 2$ or G lies within the strip between the lines

A_1B_1 and C_1D_1. Combining these two results, we conclude that either $a_G(F') = 2$ or the figure G is enclosed within each of the indicated strips—that is, is contained in the parallelogram F' (fig. 3.35). But in the latter case, G must coincide with the parallelogram F' (because G contains F').

Thus, either $a_G(F) = a_G(F') = 2$, or G coincides with the parallelogram F' (which means that G is a dilation of F); in the latter case, as we have already seen, $a_G(F) = a_G(F') = 4$. In other words, the assertion of theorem 2.2 is true for convex figures.

Now let F be a nonconvex figure. If the convex hull $\tilde{F}$ of F is a parallelogram which is a dilation of G (fig. 2.22), then $a_G(F) = 4$. If the figure $\tilde{F}$ is not a dilation of G, then by what has already been proved, $a_G(\tilde{F}) \leq 3$, and since F is contained in $\tilde{F}$, $a_G(F) \leq 3$.

The theorem is thus proved in full.

4 The Problem of Illumination

4.1. Formulation of the Problem

In this last chapter we shall consider still another problem from the theory of convex figures. The statement of this problem bears very little resemblance to those of the preceding ones, but, as we shall see, they are closely related.

Let F be a bounded convex plane figure, and let l be an arbitrary direction in the plane of this figure. We shall say that a boundary point A of F is a *point of illumination* relative to the direction l if the parallel bundle of rays having direction l "illuminates," on the boundary of F, the point A and some neighborhood of A (fig. 4.1). Note that if the line in the direction l passing through A is a supporting line for F (fig. 4.2), we do not consider A to be a point of illumination relative to the direction l. In other words, A is a point of illumination relative to the direction l if and only if the following two conditions hold:

1. The line p with direction l passing through A is not a supporting line of the figure F (that is, there are interior points of F on p).

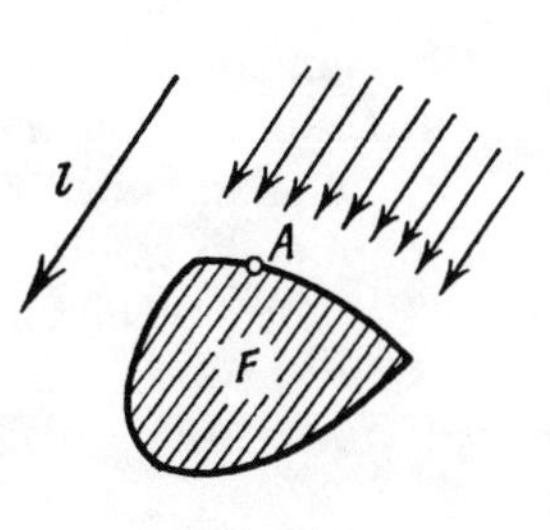

Fig. 4.1

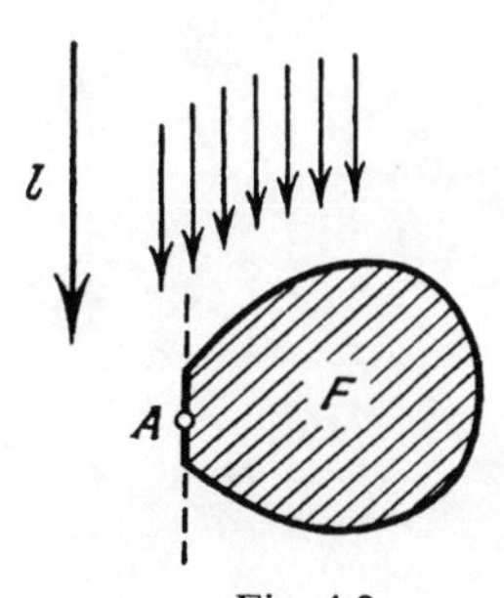

Fig. 4.2

2. The point A is the *first* point of F that we meet in moving along the line p in the direction l.

We shall also say that the directions $l_1, l_2, \ldots, l_m$ *illuminate* the entire boundary of F if each boundary point of F is a point of illumination relative to at least one of them. Finally, we shall denote by $c(F)$ the least natural number m such that there exist m directions in the plane illuminating the entire boundary of F. The problem of finding the number $c(F)$ will be called the *problem of illuminating the boundary of F*. It was posed in 1960 by V. G. Boltyanskii.

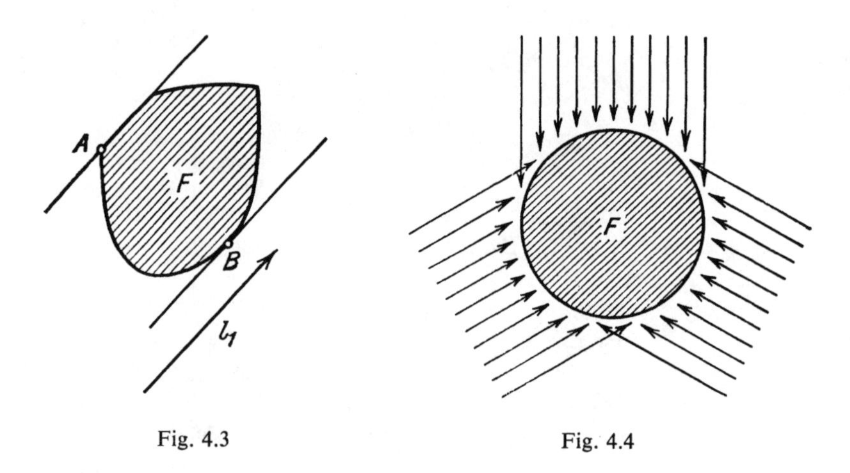

Fig. 4.3 Fig. 4.4

It is easy to prove that for any plane figure F, $c(F) \geq 3$. Indeed, let F be a bounded convex plane figure and let l_1 and l_2 be arbitrary directions. Introduce two supporting lines to F parallel to the direction l_1, and let A and B be boundary points of F lying on these supporting lines (fig. 4.3). Then neither A nor B is a point of illumination relative to the direction l_1, and the direction l_2 can illuminate no more than one of these points. Thus two directions are not enough to illuminate the entire boundary of F. For the illumination of a circle (fig. 4.4), three directions suffice. For a parallelogram (fig. 4.5), three directions are not enough (because no single direction can illuminate two vertices of parallelogram), but

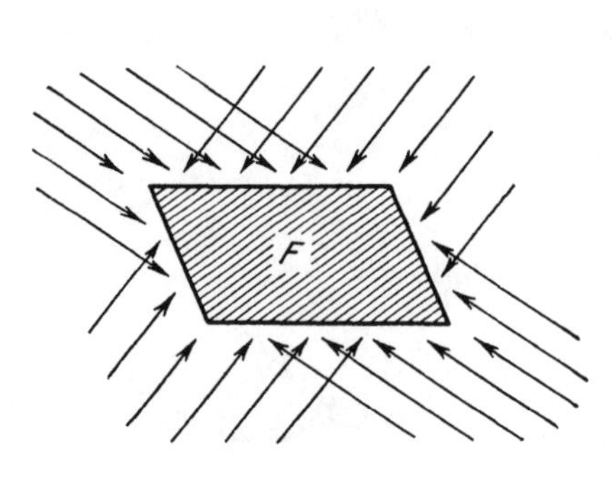

Fig. 4.5

four directions will do. Thus, for a disk, $c(F) = 3$, and for a parallelogram, $c(F) = 4$.

4.2. Solution of the Problem of Illumination

In solving the problem of illumination, we shall use the auxiliary propositions established in sec. 3.3, along with the following lemma:

LEMMA 4.1. *Let F be a convex figure, let AB be one of its major chords, and let D and E be boundary points of F lying on the same side of the chord AB. Then the entire arc DE (lying on one side of the chord AB) can be illuminated by one direction* (fig. 4.6).

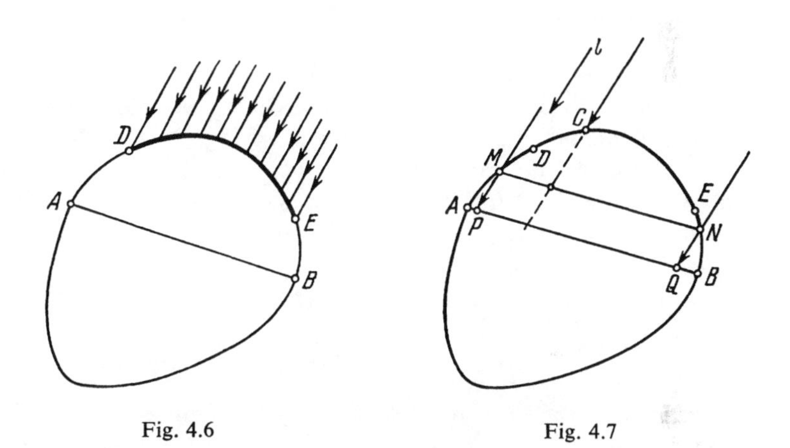

Fig. 4.6 Fig. 4.7

Proof. Let MN be a chord parallel to AB which separates the segment AB from the points D and E (fig. 4.7). Since $MN < AB$, we may pick two points P and Q in the *interior* of the segment AB such that $PQ = MN$, and thus form a parallelogram $PQNM$. Denoting by l the direction determined by the vector $MP = NQ$, it is evident that if we move along any line parallel to l passing through any point C of the arc MN, moving from C in the direction l, we cross the segment PQ, which consists entirely of interior points of the figure F. In other words, every point C of the indicated arc MN is illuminated by the direction l; that is, the entire arc DE is illuminated by l.

The lemma is thus proved.

In the problem of illumination, as in the problem of covering a figure with reduced copies of itself, the parallelogram plays a special role. Specifically, we have the following:

THEOREM 4.1. *For any bounded convex plane figure F other than a parallelogram, $c(F) = 3$; if F is a parallelogram, then $c(F) = 4$.*

Proof. The proof of this theorem is completely analogous to that of theorem 3.1. Supposing that F is not a parallelogram, theorem 3.2 implies that F has three major chords. As in the proof of theorem 3.1, we first consider the case when two major chords of F intersect on the boundary of F (fig. 3.23).

By lemma 4.1, each of the arcs KBL and DCE can be illuminated by one direction; furthermore, the points L and D can be chosen arbitrarily close to A. It is apparent that if L and D are sufficiently close to A, the remaining arc LAD can also be illuminated by one direction. Hence, in this case,

$$c(F) \leq 3 .$$

In the second case, when the three major chords AD, BE, and CH of the figure F intersect in the interior of the figure (figs. 3.24–3.26), lemma 4.1 implies that each of the arcs MBN, PAQ, and SDT can be illuminated by one direction. In this case too, then, $c(F) \leq 3$.

Finally, as has already been shown, the inequality $c(F) \geq 3$ holds for any convex figure.

Thus, if F is not a parallelogram, $c(F) = 3$, and the theorem is proved in full.

4.3. The Equivalence of the Last Two Problems

The reader has no doubt already noticed that the numbers $b(F)$ and $c(F)$ coincide for any bounded convex plane figure F (compare theorems 3.1 and 4.1).

In other words,

THEOREM 4.2. *For any bounded convex figure F,*

$$b(F) = c(F) .$$

This is true not only for plane figures but also for convex bodies in three-dimensional space (and even in higher-dimensional spaces), as proved in 1960 by V. G. Boltyanskii.

The point of proving theorem 4.2 is this: of the three theorems 3.1, 4.1, and 4.2, it is enough to prove any two; the third will be their immediate corollary. Thus one need only prove the two with the simplest proofs. In particular, if the proof of theorem 3.1 (which appears more complicated) seemed tedious to the reader and was therefore omitted, one may still obtain, upon reading the proof of theorem 4.2 and combining it with theorem 4.1, another proof of theorem 3.1. It is even more important, however, that in three-dimensional space, for which the solutions to the covering problem and the illumination problem are still unknown (see remark 16), it would suffice to solve just one of these problems, since by theorem 4.2 they are equivalent. The illumination problem has the apparent advantage of being easier to visualize.

Proof of theorem 4.2. Assume that the convex plane figure F can be covered by m reduced copies $F_1, F_2, \ldots, F_m$. Denote the center of the dilation sending F to F_i by O_i and its coefficient by k_i $(i = 1, 2, \ldots, m)$. Each of the numbers $k_1, k_2, \ldots, k_m$ is thus less than 1.

Now choose an arbitrary point A in the interior of F which is distinct from all the points $O_1, O_2, \ldots, O_m$, and denote by $l_1, l_2, \ldots, l_m$ the directions defined by the rays $O_1A, O_2A, \ldots, O_mA$. We shall prove that the directions $l_1, l_2, \ldots, l_m$ illuminate the whole boundary of F. Indeed, let B be any boundary point of F (fig. 4.8). Then B belongs to at least

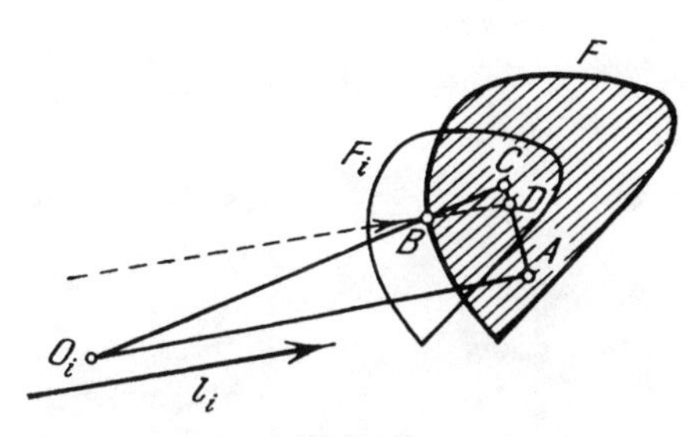

Fig. 4.8

one F_i, $1 \leq i \leq m$. Since the dilation with coefficient k_i centered at O_i maps F to F_i, there is some point C in F which is mapped into B under this dilation. Thus, $O_iB/O_iC = k_i$. Now take the point D on the segment AC satisfying $AD/AC = k_i$. The equation $O_iB/O_iC = AD/AC$ implies that BD is parallel to O_iA; that is, that the line BD is in the direction l_i. Furthermore, since the point C belongs to F, and A is an *interior* point of F, all points of the segment AC (with the possible exception of C) are interior points of F; in particular, D is an interior point.

Thus, the line BD has direction l_i and passes through the interior point D of F. It follows that B is a point of illumination relative to the direction l_i. In this manner any boundary point of F is illuminated by one of the directions $l_1, l_2, \ldots, l_m$.

We have proved that if F can be covered by m reduced copies of itself, then m directions suffice to illuminate its boundary. Thus,

$$c(F) \leq b(F) .$$

It remains to establish the reverse inequality,

$$b(F) \leq c(F) .$$

Assume that s directions $l_1', l_2', \ldots, l_s'$ illuminate the entire boundary of F. Draw two supporting lines to F in the direction l_i' (fig. 4.9), and denote by A and B the *first* points of F that we meet if we move along these lines in the direction l_i'. It is then clear that all points of the arc Δ_i with endpoints A and B (drawn with a heavy line in fig. 4.9) except the endpoints are points of illumination relative to the direction l_i'. Thus the set of points illuminated by a direction l_i' is some arc Δ_i minus its endpoints; we shall call this set the *region of illumination* relative to the direction l_i'. Since the directions $l_1', l_2', \ldots, l_s'$ illuminate the entire boundary of F, the corresponding regions of illumination cover the entire boundary of F.

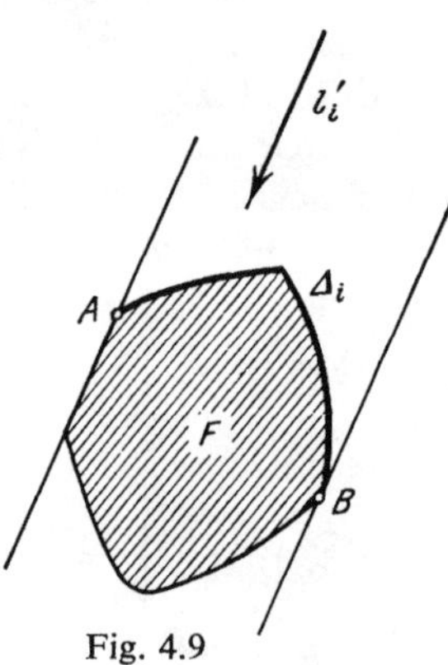

Fig. 4.9

The point A in figure 4.9 is not a point of illumination relative to the direction l_i', so it is illuminated by another direction l_j', $1 \leq j \leq s$. But then the direction l_j' illuminates all points sufficiently close to A; that is, the regions of illumination Δ_i and Δ_j *overlap*. For exactly the same reason, B, the other endpoint of the arc Δ_i, is covered by still another overlapping region of illumination Δ_k (fig. 4.10).

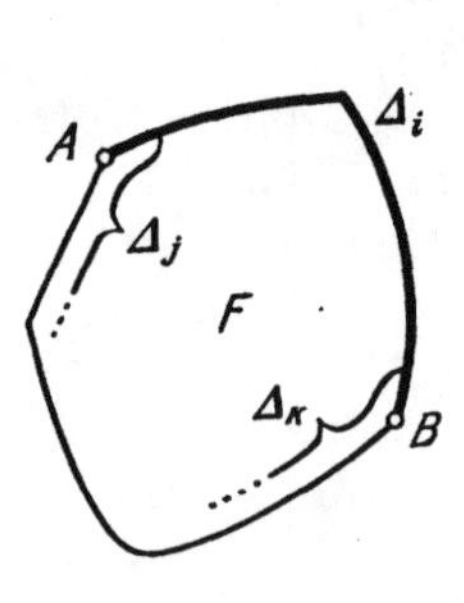

Fig. 4.10

The overlapping of the arcs $\Delta_1, \Delta_2, \ldots, \Delta_s$—the regions of illumination—implies that we can reduce each of them a little bit, with the reduced arcs still filling the entire boundary of the figure F. In other words, we can pick arcs $\Delta_1^*, \Delta_2^*, \ldots, \Delta_s^*$ that are contained (complete with endpoints) in the *interiors* of the arcs $\Delta_1, \Delta_2, \ldots, \Delta_s$, respectively (fig. 4.11), and whose union covers the entire boundary of F.

Denote the endpoints of the arc Δ_i^* by A^* and B^*. The lines through A^* and B^* in the direction l_i' must pass through interior points of F, because A^* and B^* are points of illumination relative to the direction l_i'. Denote by a and b the lengths of the chords that these lines make in F, and choose a number h_i less than both a and b. Then F completely contains the parallelogram that has A^*B^* for one side and has two sides of length h_i in the direction l_i' (fig. 4.12). It follows that any line in the direction l_i' passing through any point of the arc Δ_i^* makes a chord in F of length greater than h_i. This means that parallel translation of the arc Δ_i^* by a distance of h_i in the direction l_i' (fig. 4.13) sends Δ_i^* entirely into the interior of F. In other words, by shifting F a distance of h_i in the direction *opposite* l_i', we obtain a figure F_i^* containing the arc Δ_i^* *in its interior* (fig. 4.14). Therefore, choosing an arbitrary point O_i^* in the interior of F_i^* and applying to F_i^* a dilation with center O_i^* and coefficient $k_i^* < 1$ but sufficiently close to 1, we obtain a dilation F_i' of F_i^* (and hence of F) containing the arc Δ_i^*. Performing this construction for each $i = 1, 2, \ldots, s$ we obtain dilations $F_1', F_2', \ldots, F_s'$ of F with coefficients less than 1.

Now let O be an interior point of F. We may assume that the preceding constructions have been done in such a way that each of the figures $F_1', F_2', \ldots, F_s'$ contains O (fig. 4.15); we need only make the distances h_i sufficiently small and the coefficients k_i^* sufficiently close to 1 (recall the proof of lemma 3.1).

Finally, let G_i denote the "sector" with vertex O and arc Δ_i^* (the shaded area in fig. 4.15). Since the figure F_i' is convex and contains the arc Δ_i^* and the point O, it must contain the whole sector G_i. It follows that the union of the figures $F_1', F_2', \ldots, F_s'$ contains all of the sectors $G_1, G_2, \ldots, G_s$. But the sectors $G_1, G_2, \ldots, G_s$ clearly cover all of F (for the arcs $\Delta_1^*, \Delta_2^*, \ldots, \Delta_s^*$ cover the entire boundary of F). Therefore the figures $F_1', F_2', \ldots, F_s'$ cover the entire figure F.

We have thus proved that if the entire boundary of a figure F can be illuminated by s directions, then F can be covered by s reduced copies of itself. We therefore have the inequality

$$b(F) \leq c(F),$$

and with the inequality $c(F) \leq b(F)$ previously established, we have

$$b(F) = c(F),$$

the assertion of theorem 4.2.

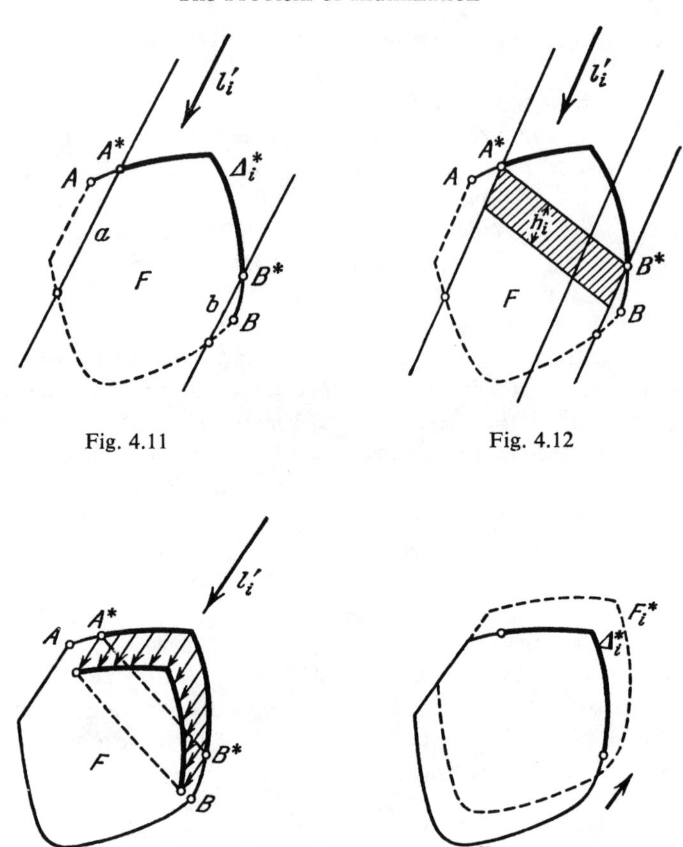

Fig. 4.11

Fig. 4.12

Fig. 4.13

Fig. 4.14

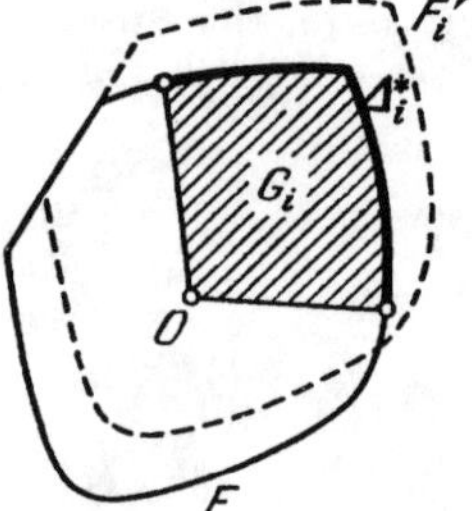

Fig. 4.15

4.4. Division and Illumination of Unbounded Convex Figures

For unbounded convex figures (see fig. 1.22), Borsuk's problem does not make sense, since such figures have no finite diameter. But the problems of illumination and of covering such a figure with reduced copies of itself are still well defined. However, a surprise awaits us at the very outset: *theorem 4.2, asserting the equality of the numbers $b(F)$ and $c(F)$, ceases to be valid for unbounded convex figures.*

A simple example of this is the convex figure F bounded by a *parabola* P. The boundary P of F can be illuminated with *one* direction; that is, $c(F) = 1$ (fig. 4.16*a*). At the same time, as we are about to see, it is

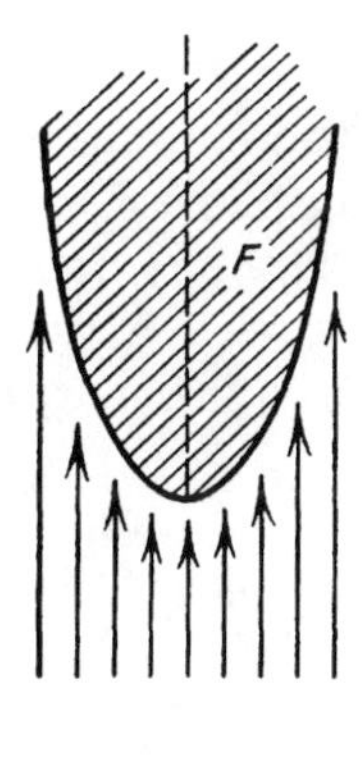

Fig. 4.16*a*

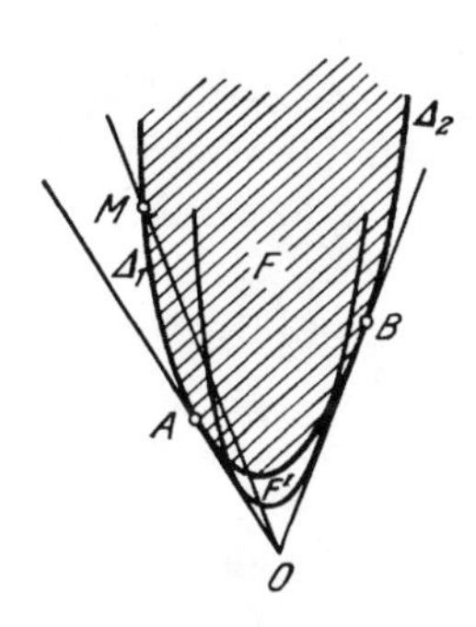

Fig. 4.16*b*

impossible to cover F with *any* finite number of reduced copies of F; that is, $b(F) = \infty$. Indeed, suppose F' is a dilation of F with coefficient $k < 1$ and center of dilation O lying outside F (fig. 4.16*b*). Draw tangents OA and OB from the point O to the parabola P. The points A and B divide P into three parts, the arc AB and two infinite arcs Δ_1 and Δ_2 terminating at A and B, respectively. It is evident that the figure F' does not contain a single point of the arcs Δ_1 and Δ_2 (for if M is any point of Δ_1 or Δ_2, then there are no points of F on the line OM beyond the point M). Thus F' can contain only a finite piece of the parabola P (some portion of the arc AB). If the center of dilation O belongs to F, then the figure F' contains at most one point of the parabola (figs. 4.16*c* and 4.16*d*). Therefore, every reduced copy of F can contain only a finite portion of P, and thus a cover of the entire figure F (which must cover P) requires infinitely many reduced copies of F; that is, $b(F) = \infty$.

Yet there do exist unbounded convex figures for which $b(F)$ is finite.

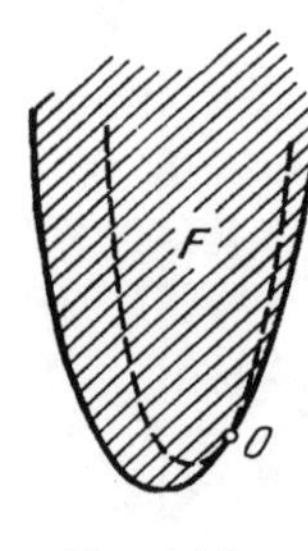

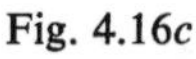
Fig. 4.16c

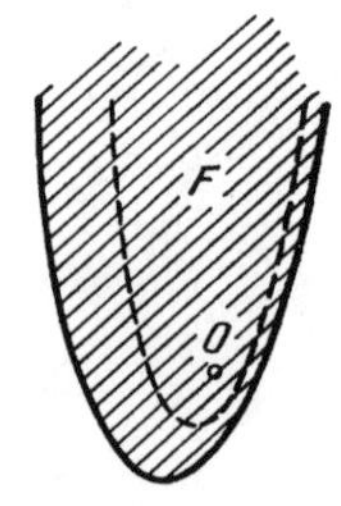

Fig. 4.16d

For example, if F is a *half-strip* (shaded in fig. 4.17a), then $b(F) = 2$; observe that in this case $c(F)$ is also 2, so that $b(F) = c(F)$.

Finally, there also exist unbounded convex figures for which $b(F)$ and $c(F)$ are both finite but are not equal. For example, if F lies within the strip between two parallel lines and the boundary of F approaches the edges of the strip as we move along one of the directions parallel to the lines defining the strip (fig. 4.17b), then it is easily shown that $b(F) = 2$ and $c(F) = 1$.

The following questions arise in connection with these considerations: *For which unbounded convex figures is the equation* $b(F) = c(F)$ *preserved? For which unbounded convex figures is* $b(F)$ *finite? Do there exist unbounded convex figures for which* $c(F) = \infty$?

These questions have been answered conclusively by P. S. Soltan and V. N. Vizitei (see remark 17). We shall state their results without proof.

First, we would point out that half of theorem 4.2 remains valid for unbounded convex figures. The first part of the proof of the theorem

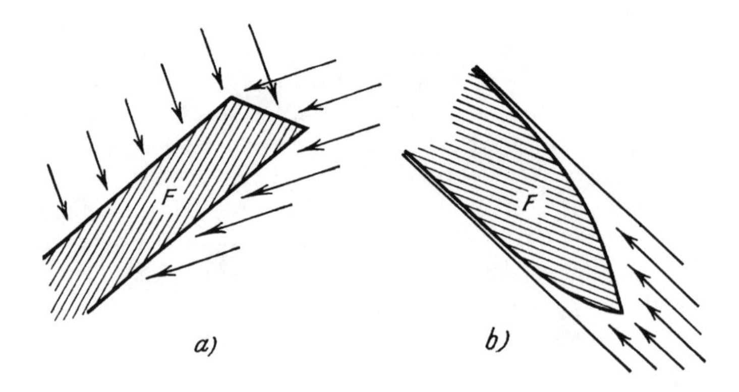

Fig. 4.17

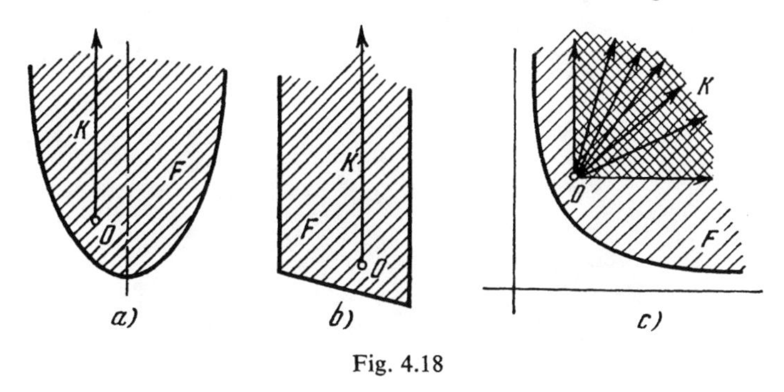

Fig. 4.18

remains wholly intact, and thus *for any unbounded convex figure F, we have the relation*

$$c(F) \leq b(F) .$$

We now formulate a theorem answering the second question. Suppose F is an unbounded convex figure. Pick an arbitrary interior point O of F and consider all rays with initial point O that lie entirely within F. The union of all such rays forms, as is easily shown, an unbounded convex figure K; K is called an *inscribed angular region* of the figure F with vertex at the point O. For example, all inscribed angular regions of a parabola (fig. 4.18*a*) or a half-strip (fig. 4.18*b*) consist of a single ray; for the interior region of a branch of a hyperbola, an inscribed angular region is shown in figure 4.18*c*. (Observe that if one takes as the vertex any other interior point of F in place of O, the inscribed angular region is not changed in shape or size, but only translated.)

P. S. Soltan calls an unbounded convex figure F *almost conical* if there exists a (finite) number r such that some inscribed angular region K comes within a distance r of all points of F. Thus, the figures shown in figures 4.18*b* and 4.18*c* are almost conical, but the one in figure 4.18*a* is not, since points of the parabola lie at arbitrarily large distances from any vertical axis.

THEOREM 4.3. *For an unbounded convex figure F, b(F) is finite if and only if F is almost conical.*

Moreover, for almost conical figures F, the function $b(F)$ can take on only the values 1 and 2. Specifically, let F be a two-dimensional unbounded convex almost conical figure not containing any line; if its inscribed angular regions are rays, then $b(F) = 2$; otherwise $b(F) = 1$. Finally, if a two-dimensional convex figure contains a line, then it must either be a strip, a half-plane, or a plane; in these cases $b(F)$ takes the values 2, 1, and 1, respectively. With this the determination of the values of $b(F)$ for all unbounded convex plane figures is complete.

Remarks

1. In order to assure that each figure F considered will have a well-defined diameter, it is necessary to assume that every such figure is both *closed* (contains all its boundary points) and *bounded* (can be enclosed in some suitably large disk). If not for the latter condition, arbitrarily large distances between points of F could be attained. If not for the former, we could consider such objects as an *open* disk of diameter d (that is, a disk without the points on its circumference). In this case, the *supremum*, or *least upper bound*, of the distances between pairs of points would be d, yet there would be no pair of points separated by a distance of exactly d.

But if F *is* a closed bounded set (in the plane or in a Euclidean space of an arbitrary number of dimensions), there will necessarily be some two points A and B of F at maximal distance from one another. Indeed, for any two points M and N of the set F, let $\rho(M, N)$ be the distance between them. The function $\rho(M, N)$ is *continuous* [in the ordered point (M, N)]. But any continuous function (in this case, of the two variables M and N) whose arguments vary over closed bounded sets necessarily achieves a maximum (and a minimum) value. Thus, there exist two points A and B of F such that $\rho(A, B) \geq \rho(M, N)$ for any points M and N of F. The distance $d = \rho(A, B)$ between such a pair of points is the diameter of the set F.

2. Indeed, suppose A and B are two points of the polygon F. If A is not a vertex of the polygon, then there is a segment CD containing A in its interior which is entirely contained in F (figs. R1*a* and R1*b*). Without loss of generality, label the points so that $\angle BAC \geq \angle BAD$. Then $\angle BAC \geq 90°$, so that $\angle BAC > \angle BCA$, and therefore $BC > AB$. AB is consequently not a diameter of F. Thus, if either of the points A or B is not a vertex of the polygon, then AB is not a diameter of the polygon. This implies that the diameter of the polygon will be the distance between two of its *vertices* (specifically, the two vertices which are furthest apart).

3. Here we are concerned with dividing a figure into pieces having well-defined diameters. With the previous remark in mind, we shall assume that

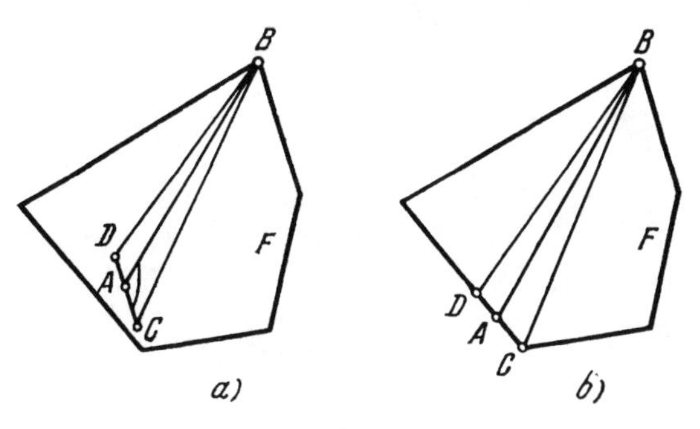

Fig. R1

the pieces into which the figure is divided are themselves closed sets. Hence the proposition in the text could be more precisely stated as follows: *if the disk F of diameter d is somehow represented as the union of two of its closed subsets, then at least one of these subsets has the same diameter d.* The argument presented on page 3 does not, of course, give a complete proof of this assertion. A rigorous proof goes like this: denote the two closed subsets by H_1 and H_2 (so that the *union* $H_1 \cup H_2$ gives the whole disk F). The points of the set H_1 that lie on the circumference of F constitute a certain set K_1; the set K_2 is defined analogously. Thus, the circle that is the circumference of F can be represented as the union of two of its closed subsets, K_1 and K_2. If one of these sets is empty (does not contain any points), then the other is identical with the whole circle; therefore this K_i and the larger set H_i have diameter d. If both sets K_1 and K_2 are non-empty, then they necessarily have some common point A (for the circle is *connected* and therefore cannot be represented as the union of two disjoint closed subsets). Let B denote the point diametrically opposite A, and suppose, without loss of generality, that B belongs to the set K_2. Then K_2 contains both of the points A and B. Hence K_2, and thus H_2, has diameter d. In either case, then, at least one of the sets H_1 and H_2 has diameter d.

4. Let us make one more remark about the "division" of figures into pieces. The word "division" can be understood to mean that the figure F is represented as the union of several of its subsets:

$$F = H_1 \cup H_2 \cup \cdots \cup H_m .$$

In this case mathematicians usually say that the sets $H_1, H_2, \ldots, H_m$ form a *covering* of the figure F. But it is more natural for the term "division" to mean that the closed sets $H_1, H_2, \ldots, H_m$ not only form a covering of F, but also do not *overlap*: that is, no two of them have any common interior points.

One easily recognizes that the essential content of the problem of dividing figures into pieces of smaller diameter does not depend on which way one interprets the term "division." Indeed, if the figure F is represented as the union of several of its closed subsets

$$F = H_1 \cup H_2 \cup \cdots \cup H_m$$

(possibly overlapping), then we can, without increasing the diameters of the pieces, "adjust" these pieces so that they will not overlap. For this, observe that the sets[1]

$$\begin{aligned}
&H_1 \backslash (H_2 \cup H_3 \cup \cdots \cup H_m), \\
&H_2 \backslash (H_3 \cup \cdots \cup H_m), \\
&\cdots\cdots\cdots\cdots\cdots\cdots, \\
&H_{m-2} \backslash (H_{m-1} \cup H_m), \\
&H_{m-1} \backslash H_m, \\
&H_m,
\end{aligned}$$

constitute a covering of F, and that no two of them intersect. It is true that these sets may not turn out to be closed. However, the *closures*[2] of these sets, that is, the sets

$$\begin{aligned}
H_1' &= \overline{H_1 \backslash (H_2 \cup H_3 \cup \cdots \cup H_m)}, \\
H_2' &= \overline{H_2 \backslash (H_3 \cup \cdots \cup H_m)}, \\
&\cdots\cdots\cdots\cdots\cdots\cdots \\
H_{m-1}' &= \overline{H_{m-1} \backslash H_m}, \\
H_m' &= H_m,
\end{aligned}$$

will be closed subsets of F with pairwise disjoint interiors (that is, nonoverlapping closed sets), and will cover F.

Thus, from an arbitrary covering $(H_1, H_2, \ldots, H_m)$ of F by closed subsets we obtained a cover $(H_1', H_2', \ldots, H_m')$ consisting of nonoverlapping closed sets. The diameters of the pieces have not, of course, been increased (as H_i' is contained in H_i).

5. The problem of dividing figures into pieces of smaller diameter [that is, the problem of determining what values $a(F)$ can assume] can be studied not only for *plane* figures, but also for *solid* bodies or even for sets in *n-dimensional Euclidean space*. It was for n-dimensional bodies that

1. The symbol $A \backslash B$ denotes the set of points of A that are not in B.

2. The *closure* of a set is the union of that set with its boundary points (and thus is the "smallest" closed set containing the given set).

Borsuk formulated his problem. He found a complete solution only for plane figures (Borsuk's theorem is presented in sec. 1.3 of this book). Borsuk also proved that for an n-dimensional ball the value of $a(F)$ is $n + 1$. [In particular, $a(F) = 4$ for the three-dimensional sphere.] In connection with this he proposed the following problem: *to prove that for any n-dimensional body F the inequality* $a(F) \leq n + 1$ *holds.* Borsuk's elegant solution of the case $n = 2$ and the charming simplicity of the statement of the problem attracted the attention of many mathematicians. But for $n > 2$ the problem turned out to be considerably more complicated. Only in 1955 (twenty-five years after the problem was posed) did the English mathematician Eggleston succeed in solving the problem for $n = 3$ by proving that $a(F) \leq 4$ for any three-dimensional body. Proofs simpler than that of Eggleston were found two years later by the American mathematician Grunbaum and the Hungarian mathematician Heppes. Borsuk's problem for $n > 3$ is still unsolved. The reader who is interested in the higher-dimensional cases may read more on this subject in the book *Problems and Theorems in Combinatorial Geometry* by the same authors.

6. The argument in the text (in which the line l "approaches" the figure) does not, of course, provide a rigorous proof of the existence of the supporting line l_1. One way to obtain a rigorous proof is the following: Draw a line l not intersecting F, and a second line m perpendicular to l. Let the lines l and m be coordinate axes (fig. R2), and let $y(A)$ denote the *ordinate* of any point A of F (measured along the line m). Thus a function y is defined on the figure F, and this function is continuous [for the difference $|y(A) - y(A')|$ does not exceed the distance AA']. But a continuous function defined on a closed bounded set F achieves maximum and minimum values. In other words, there exist points M_1 and M_2 of F such that

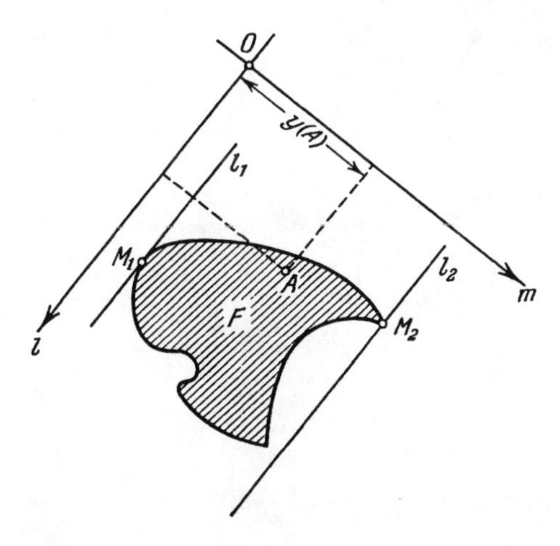

Fig. R2

$y(M_1) \leq y(A) \leq y(M_2)$ for every point A of F. But this means that if we draw lines through M_1 and M_2 parallel to the first axis l, all of the figure F will be contained in the strip between these lines. The lines l_1 and l_2 that pass through the points M_1 and M_2 parallel to l are thus the two supporting lines to F referred to in the text (see fig. 1.14).

7. We will show, as an example, that the point A depends continuously on the direction of the line l_1. Suppose the line l_1 has been turned by some angle α (fig. R3). The positions of the lines l_1, l_2, m_1, and m_2 at this moment will be denoted by l_1', l_2', m_1', and m_2'. Draw lines m_A and m_B through the points A and B so as to form an angle of α with m_1 (that is, parallel to m_1'). Since the line m_A cuts across F and the line m_B misses F completely (except possibly for the point B), the supporting line m_1' must be located *between* m_A and m_B (inclusively). Similarly, if we draw lines l_A and l_D through A and D, forming an angle α with the line l_2 (that is, parallel to l_2'), we find that the supporting line l_2' is located between l_A and l_D (inclusively). Hence the point A', where the supporting lines m_1' and l_2' cross, is located in the parallelogram formed by the line m_A, m_B, l_A, and l_D. But the dimensions of this parallelogram (shaded in fig. R3) can be made arbitrarily small, provided only that the angle α is sufficiently small. Thus the point A' is

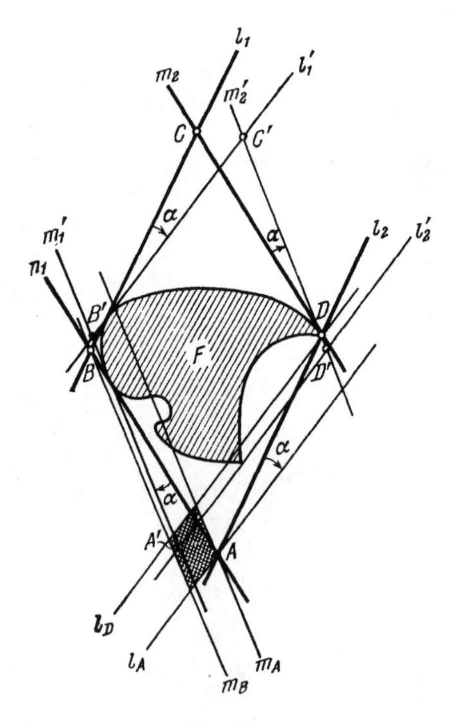

Fig. R3

arbitrarily close to A if the angle α is sufficiently small. This proves the continuity of A as a function of the direction of the line l_1.

8. Indeed, suppose M is a collection of convex figures, and let Φ be the intersection of all these figures (that is, the set of points belonging to *all* of the figures in M). We shall prove that the figure Φ is convex. Let A and B be two points of Φ, and let F be an arbitrary figure from the collection M. Then the points A and B belong to the figure F. Since F is convex, it follows that the entire segment AB is contained in F. Thus *every* figure F of the collection M contains the segment AB, and therefore the figure Φ contains the segment AB. This shows that Φ is convex.

9. F^*, an intersection of closed disks, is closed and bounded, and thus has a diameter.

10. In high school geometry, the concept of *area* is defined only for the simplest figures—for example, polygons and circular disks. In fact, however, the notion of area can be defined for considerably larger classes of figures. In particular, one may speak of the *area* of any convex bounded figure. We shall not present a definition of area for convex figures here. (See the articles "Area and Volume" and "Convex Figures and Solids" in vol. 5 of the [Russian] *Encyclopedia of Elementary Mathematics*.) What is important to us here is the fact that to each bounded figure F there corresponds some positive number $s(F)$, the area of the figure, and that if a bounded convex figure F' contains the figure F and has interior points not belonging to F, then $s(F') > s(F)$.

11. H has been defined as the union of an *increasing sequence* of convex figures $H_0, H_1, \ldots, H_n, \ldots$ (that is, of a sequence in which each figure contains the preceding one). It is proved in the text that the figure H is convex and has diameter d. Note, however, that the union of an increasing sequence of convex figures (even closed convex figures) might *not* be a closed convex figure. For example, let $F_1, F_2, \ldots, F_n, \ldots$ be an increasing sequence of (closed) disks, all internally tangent to one another at one point M, and let the radius of each disk F_n be $1-(\frac{1}{2})^n$ (fig. R4). Then the union of this increasing sequence of convex figures will be a disk of radius 1 which includes the point M but does not include the other boundary points. In other words, this union is a *nonclosed* convex figure.

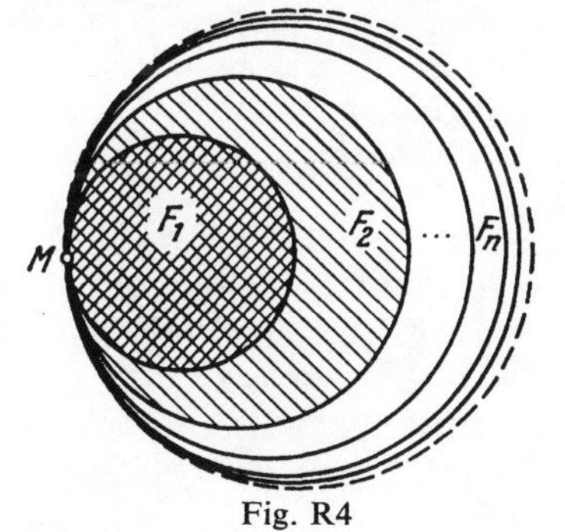

Fig. R4

Since our policy is to consider only *closed* convex figures, the figure H ought, strictly speaking, to be defined in the following way: take the union of all the figures H_0, H_1, $H_2, \ldots, H_n, \ldots$, and add to the convex figure thus obtained all of its boundary points (that is, take its closure). But this does not change the essence of the matter.

12. Observe that theorem 1.3 cannot immediately be generalized to the case of solid (three-dimensional) bodies. For example, for a *regular tetrahedron* F with edge d, we readily see that $a(F) = 4$; that is, $a(F)$ takes on its maximal value. Nevertheless, the solid F has more than one completion to a solid of constant width d (see pp. 103–4 of the book by I. M. Yaglom and V. G. Boltyanskii mentioned in chap. 1, n. 1). The property $a(F) = 4$ for solid bodies involves more subtle conditions than uniqueness of completion to a solid of constant width d.

13. Indeed, suppose l and l' are two parallel supporting lines to the figure F (fig. R5), and let A and B be points of contact of these lines with F.

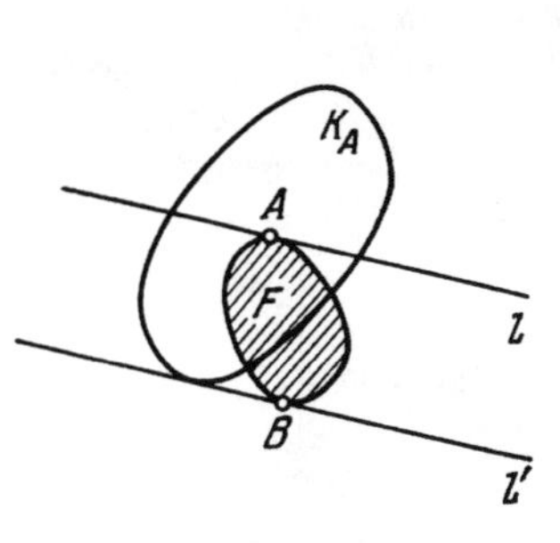

Fig. R5

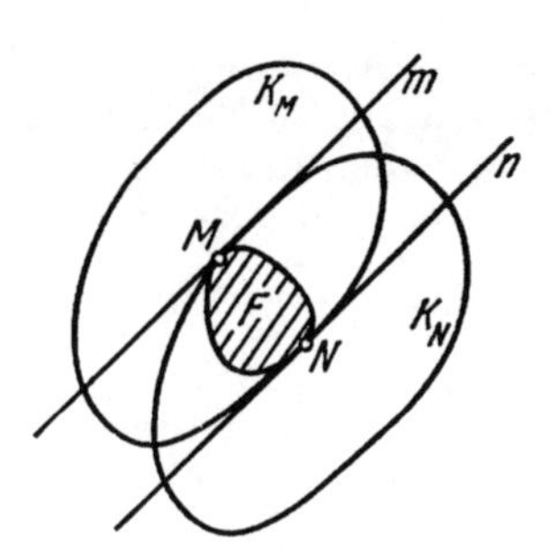

Fig. R6

Then the Minkowski distance h between the lines l and l', the width of F in the direction l, cannot exceed the distance $d_G(A, B)$, since A is a point of l and B is a point of l'. Thus the width of F in any direction does not exceed its diameter d, and so the greatest width does not exceed d.

It remains to be proved that there is some direction in which the width of F *equals* d. Let M and N be two points of F separated by the maximal distance d. Then the disk K_M of radius d centered at M completely contains F. The disk K_N of radius d centered at N likewise contains F (fig. R6). Draw a supporting line n to the disk K_M passing through the point N. Then the line m which passes through the point M parallel to the line n is a supporting line to the disk K_N. It is evident that the figure F is entirely contained within the strip between the lines m and n, so that m and n are supporting lines to F. It is clear from the construction that the distance between the lines m and n (that is, the width of F in the direction m) is equal to the radius of the disk K_M, namely d.

14. Let us prove this using the concept of the *sum* of convex figures (see sec. 4 of the book by I. M. Yaglom and V. G. Boltyanskii mentioned in chap. 1, n. 1).

Let F be an arbitrary convex figure, and let F' be the figure symmetric to F with respect to some origin O. Let G denote the sum $F + F'$. Then G is a centrally symmetric convex figure and may be taken as the unit disk for a Minkowski plane. It is clear that the width of F in any direction l is equal to the width of F' in the same direction (fig. R7). The width of G in an arbitrary direction is consequently twice the width of F in that direction.

But the width of G in any direction is 2 (since G is a unit disk). Thus, the width of F in any direction is 1; F is (in the Minkowski plane defined by the unit disk G) a figure of constant width 1.

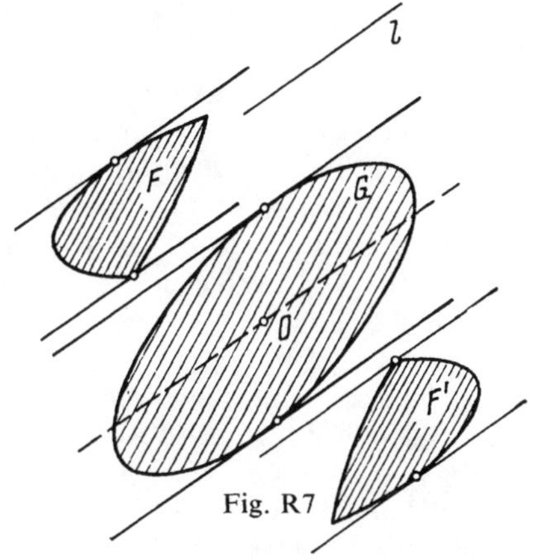

Fig. R7

Of course, if we replace the unit disk $G = F + F'$ with any dilation of G, F will also be a figure of constant width in the new Minkowski plane.

To prove uniqueness, suppose F is a figure of constant width in some Minkowski geometry. Then F' is clearly a figure of constant width, as is $F + F'$. But the figure $F + F'$ is centrally symmetric, and, as is easily seen, a centrally symmetric convex figure has constant width only if it is a disk in the given Minkowski plane.

Thus we have proved the following theorem:

THEOREM R1. *A figure F is a figure of constant width in the Minkowski plane with unit disk G if and only if the sum $F + F'$ of F and the figure F' centrally symmetric to F is a disk in this geometry.*

15. Let F be a convex bounded figure, and let G be a figure contained in F (that is, a closed subset of F). Consider all dilations of F with coefficients less than or equal to 1 containing the figure G. Denote by k_0 the infimum (greatest lower bound) of all the coefficients of dilation for such figures. If $k_0 = 1$, the bulk of the piece G is 1 (because there is no dilation of F with coefficient less than 1 containing G, but there is a dilation of F with coefficient exactly 1, F itself containing G). If $k_0 < 1$, we may choose a sequence $F_1, F_2, \ldots, F_q, \ldots$ of dilations of F with centers of dilation $O_1, O_2, \ldots, O_q, \ldots$, respectively, and with coefficients of dilation $k_1, k_2, \ldots, k_q, \ldots$ such that each F_i contains G and such that $\lim_{q\to\infty} k_q = k_0$. The sequence can, of course, be chosen so that the coefficients decrease monotonically: $1 > k_1 > k_2 > \cdots > k_q > \cdots > k_0$.

It is easily shown that all of the points $O_1, O_2, \ldots, O_q, \ldots$ are situated at distances no greater than $d/(1 - k_1)$ from the figure F (where d is the diameter of F). Indeed, suppose the distance between O_q and F is greater than $d/(1 - k_1)$. Then $O_qA > d/(1 - k_1)$ for every point A of F. Under the dilation with center O_q and coefficient k_q, the point A is mapped to a point A' on the line O_qA satisfying the equation $O_qA' = k_q \cdot O_qA$. Therefore $AA' = (1 - k_q) \cdot O_qA > (1 - k_1) \cdot O_qA > d$. Since each point A of F is thus moved by a distance greater than d under this dilation, each is sent to a point A' which does *not* belong to F. In other words, the figure F_q to which F is mapped has no points in common with F. But this contradicts the assumption that F_q contains the piece G.

Thus there is an upper bound on the distances from F to all of the points $O_1, O_2, \ldots, O_q, \ldots$. The sequence $O_1, O_2, \ldots, O_q, \ldots$ therefore has at least one limit point. Without loss of generality we may assume (passing to a subsequence if necessary) that the sequence $O_1, O_2, \ldots, O_q, \ldots$ *converges* to some point O_0; that is, $\lim_{q \to \infty} O_q = O_0$.

It is clear that the dilation F_0 of F with center O_0 and coefficient k_0 contains G (since $\lim_{q \to \infty} k_q = k_0$ and $\lim_{q \to \infty} O_q = O_0$). Thus there actually exists a dilation F_0 of F with coefficient k_0 containing G; yet no dilation of F with coefficient less than k_0 can completely contain G (by the definition of infimum). It is thus established that the concept of bulk is defined for any piece G of the figure F.

16. The problem of illumination is considerably more interesting and more difficult for three-dimensional and n-dimensional bodies. More information on the results stated here and on the unsolved problems can be found in the book by V. G. Boltyanskii and I. Ts. Gokhberg mentioned in the preface. We wish only to point out that for bounded three-dimensional convex solids (and even for three-dimensional polyhedra) it is still unknown whether the inequality $c(F) \leq 8$ always holds.

17. The problems of covering and illumination of unbounded figures become especially interesting in the three-dimensional and n-dimensional cases. P. S. Soltan has proved that theorem 4.3 is valid for figures in any number of dimensions. In particular, Soltan completely solved the problem of covering unbounded three-dimensional solids by "reduced copies"; the function $b(F)$ in these cases can assume only the values 1, 2, 3, 4, and ∞.